THE CYCLICAL SERPENT
Prospects for an Ever-Repeating Universe

THE CYCLICAL SERPENT
Prospects for an Ever-Repeating Universe

PAUL HALPERN

Foreword by
Andrei Linde

All drawings courtesy of Felicia Hurewitz

PLENUM PRESS • NEW YORK AND LONDON

Library of Congress Cataloging-in-Publication Data

Halpern, Paul, 1961-
　　The cyclical serpent : prospects for an ever-repeating universe / Paul Halpern ; foreword by Andrei Linde ; all drawings courtesy of Felicia Hurewitz.
　　　　p.　　cm.
　　Includes bibliographical references and index.
　　ISBN 0-306-44923-4
　　1. Cosmology.　　I. Title.
QB981.H25　1995
523.1--dc20
　　　　　　　　　　　　　　　　　　　　　　　　　　　　　　95-1806
　　　　　　　　　　　　　　　　　　　　　　　　　　　　　　CIP

Some parts of the Foreword are based on the materials contained in "The Self-Reproducing Inflationary Universe," *Scientific American*, Vol. 271, No. 5 (November 1994): 48–55 and in *Particle Physics and Inflationary Cosmology* (Chur, Switzerland: Harwood Academic Publishers, 1990), both of which were written by Andrei Linde.

ISBN 0-306-44923-4

© 1995 Paul Halpern
Plenum Press is a Division of Plenum Publishing Corporation
233 Spring Street, New York, N.Y. 10013-1578

10 9 8 7 6 5 4 3 2 1

All rights reserved

No part of this book may be reproduced, stored in a retrieval system, or transmitted in any form or by any means, electronic, mechanical, photocopying, microfilming, recording, or otherwise, without written permission from the Publisher

Printed in the United States of America

ACKNOWLEDGMENTS

Thanks to Robert Marande, Nancy Cunningham, Elizabeth Bressi-Stoppe, Charles Gibley, Philip Gerbino, and Allen Misher for their strong support, to Durai Sabapathi for his insightful comments on Hinduism, and to other members of faculty and staff of the Philadelphia College of Pharmacy and Science. I also appreciate the encouragement of my family and friends for this project, including my parents, Stanley and Bernice, and my brothers, Kenneth, Alan and Richard, as well as Elana Doering, Michael Erlich, Fred Schuepfer, Joseph and Arlene Finston, Janet Gala, Jennifer and Fred Schwartz, Simone Zelitch, Scott Veggeberg, and Carolyn Brodbeck. Thanks to Andrei Linde for numerous helpful comments. My agent, John Ware, deserves a special word of thanks for his enthusiastic work on this project, as does my editor, Linda Greenspan Regan, for her many helpful suggestions. And most of all, heartfelt thanks to my new wife, Felicia Hurewitz, for her loving support.

Acknowledgments

The graphic art in this book, aside from the photographs, was rendered by Felicia Hurewitz.

FOREWORD

By Andrei Linde

Where did our Universe come from? What is its fate? What is the origin of life? What will happen to us in the future?

For thousands of years people have been trying to answer these questions. They came up with a variety of completely different ideas. Western science originally considered the Universe as being static and perhaps even finite. Meanwhile, Indian culture gave rise to the idea of an eternally changing Universe, coming through many periods of formation and destruction.

The 20th century brought to life new theories describing our Universe, and new experimental evidence supporting them. First of all, it was realized that our Universe is not static. It changes in time in accordance with Einstein's theory of gravity. A new picture of the world emerged, which was called the Big Bang theory. This theory asserted that at some initial moment of time the whole Universe was created from "nothing" as a huge rapidly growing ball of fire. As the Universe expanded, this fire subsided. Its remnants still surround us in the form of the cosmic microwave background radiation with a very small temperature (2.7 kelvins).

Creation of this new cosmological paradigm was extremely painful. Even Einstein originally did not want to accept the idea that the Universe changed in time; he tried to modify his own equations in order to come back to the old picture of a static Universe. However, the theory of the Big Bang was in a very good agreement with observational data, and gradually everybody agreed that it must be correct. It seemed that we have finally discovered the ultimate theory describing our cosmic home.

Unfortunately, this attitude was too optimistic. By the end of 1970's it was understood that it is very difficult to make cosmology consistent with the new theories of elementary particles. Then everybody suddenly realized that the standard Big Bang theory is plagued by many different problems. What was before the Big Bang? Why is the Universe so big? Why is its geometry so similar to the geometry of a flat table where parallel lines do not intersect? Why is the Universe so homogeneous? If it must be homogeneous, then how have galaxies been created? How could all the different parts of the Universe synchronize the beginning of their expansion?

All these problems (and others which are too technical to be discussed here) are extremely difficult. Fortunately, fifteen years ago it was realized that many of them can be resolved in the context of a relatively simple theory, the theory of inflationary Universe. According to this theory, the Universe at the very early stages of its evolution came through the stage of inflation, exponentially rapid expansion in a kind of unstable vacuum-like state (a state with large energy density, but without elementary particles). This stage could be very short, but the Universe within this time became exponentially large. At the end of inflation the vacuum-like state decayed, the Universe became hot, and its subsequent evolution could be described by the standard Big Bang theory.

This theory originally was considered as a particular version of the Big Bang scenario. However, at the present time inflationary theory seems to be much more general than the original Big Bang theory. During the last fifteen years, inflationary cosmology has encompassed the theory of the Big Bang, in the same sense as Einstein's general theory of relativity encompassed the theory of gravity developed by Newton. Without understanding of this theory it is difficult to appreciate many other ideas discussed in this book. Therefore we will say here a few words about inflation.

There were several different attempts to develop inflationary theory. Unfortunately, because of political barriers between Russia and the United States, the history of this theory is only partially known to American readers.

The first realistic version of inflationary cosmology was proposed in 1979 by Alexei Starobinsky of the L. D. Landau Institute of Theoretical Physics in Moscow. For two years his model remained the main topic of discussion at all conferences on cosmology in the USSR. However, his model was rather complicated, being based on the theory of anomalies in quantum gravity, and it did not really address the problem of initial conditions for inflation.

In 1981 Alan H. Guth of the Massachusetts Institute of Technology suggested a much simpler version of inflationary theory, based on the theory of cosmological phase transitions with supercooling. This theory was proposed in 1972 by David A. Kirzhnits and me at the P. N. Lebedev Physics Institute in Moscow. The model suggested by Guth (which now is called "old inflation") had a very clear physical motivation, and it immediately became very popular. Unfortunately, unlike the Starobinsky model, the old inflationary theory did not work. After investigating his model for a year, Guth finally renounced it in a paper he co-authored with Erick J. Weinberg of Columbia University.

In 1982 I introduced the new inflationary universe scenario, which later was also discovered by Andreas Albrecht and Paul J. Steinhardt of the University of Pennsylvania. This scenario was free of the main problems of Guth's model, but it was still not very realistic.

Most of the inflationary models which exist at present are based on the idea of chaotic inflation which I have proposed in 1983. Chaotic inflation scenario does not require complicated quantum gravity effects or phase transitions with supercooling. It does not require even the standard assumption that the Universe originally was hot. The simplest version of this scenario describes the Universe filled by a so-called massive scalar field. Such fields appear in most of the theories of elementary particles developed during the last 25 years. It can be shown that if the scalar field originally was sufficiently large and homogeneous, it could change its value only very slowly. Therefore, for a long time its energy density remained almost constant. According to Einstein's theory of gravity, the Universe filled by matter with a constant energy density should expand exponentially, which corresponds to the stage of inflation.

The main idea of the chaotic inflation scenario is so simple that it is hard to understand why it was not discov-

Foreword

ered 50 years ago. Meanwhile its consequences are very dramatic. In the simplest version of this theory the Universe during inflation could expand approximately $10^{10^{12}}$ times. This number is model-dependent, but it is huge in all realistic inflationary models. Such an expansion makes the Universe extremely large and stretches away all its inhomogeneities. Geometric properties of our space become similar to the properties of an almost flat surface of a huge inflating balloon. This explains why our Universe is so big, why it is so homogeneous and isotropic, and why its geometric properties are so close to the properties of a flat space.

Not only the Universe, but even small quantum fluctuations of various fields rapidly grow during inflation. Such fluctuations always exist in vacuum, but typically they are microscopically small and short-lived. However, quantum fluctuations amplified during inflation may become extremely large. These fluctuations later gave rise to galaxies, and also to small perturbations in the 2.7 kelvin temperature of the microwave background radiation. In 1992 these perturbations were found by the Cosmic Background Explorer (COBE) satellite, a finding later confirmed by several other experiments.

There is one peculiar property of the new theory which makes it especially relevant to the discussion contained in this book. In certain cases quantum fluctuations amplified during the rapid expansion of the Universe may considerably increase the value of the scalar field which drives inflation. The probability of such events is very small, but those rare parts of the Universe where it happens begin expanding with even greater speed. This creates a lot of new space where inflation may occur, and where large quantum fluctuations become possible. As a result, the Universe enters a stationary regime of self-reproduction: Inflationary

Universe permanently produces new inflationary domains, which in turn produce new inflationary domains. Such a Universe looks not like a single expanding fireball created in the Big Bang, but like a huge self-reproducing fractal, like a tree consisting of inflationary balls producing new and new balls, of all possible types.

According to many inflationary models, quantum fluctuations at the stage of inflation could be so large that they could change properties of the vacuum state in different parts of the Universe. As a result, the Universe becomes divided into many exponentially large parts where the laws of low-energy physics and even dimensionality of space-time may be different.

If this theory is correct, we are witnessing the most dramatic change in our picture of the world. The Universe becomes immortal. Different parts of the Universe may appear and disappear again, but the Universe as a whole will exist forever. This does not mean that our own civilization can live for an indefinitely long time. Life in our part of the Universe may eventually disappear, but even if this happens, there will be infinitely many other parts of the Universe where life will appear again and again, in all its possible forms.

The new vision of the world may be considered as a synthesis of ideas elaborated by many different cultures. On a small scale the Universe looks like a single spherically symmetric expanding ball of matter created in the Big Bang. On a greater scale we see new parts of the Universe appearing and disappearing again. And if we try to investigate the global structure of the Universe—the distribution of these new balls, the properties of matter inside them—then we may see that these properties all over the whole Universe on average do not depend on time. Just like a river which brings new water but still looks the same, the Universe changes all the time, but in a certain sense it remains stationary.

If this is the case, then we may be forced to abandon the standard assumption that physics alone can give a complete explanation to the properties of the world around us. This assumption was based in part on the idea that the Universe is everywhere the same. This belief was called the "cosmological principle." If the Universe is everywhere the same, then there should be some physical reason why it appeared in this particular state. Thus one should find a theory which explains why our Universe must be the way it is, why the proton must be 2000 times heavier than the electron, why our space-time must be four-dimensional, etc.

The cosmological principle was based on the observational fact that our part of the Universe on a very large scale is almost exactly uniform. At present we know only one physical mechanism which explains why the Universe around us is so uniform. It is given by inflationary cosmology. However, simultaneously with explaining why our part of the Universe is homogeneous, this theory predicts that on a much greater scale the Universe must be absolutely inhomogeneous, that there are some places in the Universe where inflation still goes on, and that the Universe after inflation may become divided into many exponentially large parts with different properties.

Thus, inflationary theory effectively eliminates the cosmological principle. Instead of insisting that the Universe everywhere must be the same, we should examine all those exponentially large parts of the Universe which can emerge after inflation. Then we should take into account that *we* cannot live in those parts where the electrons are too heavy, where the gravity forces are much stronger, or where the number of dimensions of space-time is different from four.

Thus we can get a partial explanation of the properties of our world by eliminating from our consideration all those parts of inflationary Universe where we cannot live.

This philosophy is based on the so-called anthropic principle. This principle was not particularly popular among physicists. It was based on a hidden assumption that the Universe was created many times, one after another, until the final success. But who was doing this job, what was the purpose, and why it was necessary for him to work so hard? There was absolutely no need to create the Universe which was almost perfectly homogeneous far away from us. It would be quite sufficient to create a small island of homogeneity in a vicinity of a solar system.

The new cosmological theory gives us a possibility to answer these questions. Inflationary Universe recreates itself in all its possible forms, and when it does so, it creates islands of homogeneity of a typical size much greater than the size of the part of the Universe which we can see now.

Thus, the anthropic principle becomes an important part of the scientific approach to an explanation of our world. There is nothing mysterious about it; we are just trying to use our knowledge of the inhabitants of our cosmic home to explain the properties of the small part of the Universe where they can live. However, one may try to go in an opposite direction as well. We may try to use our investigation of the properties of the Universe to understand something about ourselves.

For example, several years ago we thought that the whole Universe was born about 15 billion years ago, and if the Universe is closed then it should eventually collapse and disappear. However, now inflationary theory teaches us that the Universe as a whole never dies. Moreover, there is an interesting possibility presently under investigation. It may

Foreword

happen that when the density of a collapsing part of the Universe becomes sufficiently large, it begins to expand again, though not in our space, but in some other space which is in a certain sense dual to ours.

It is hard to say whether this example can teach us any lesson concerning our own fate. We know that each of us was born some time ago. We also know that each of us is going to die, and the whole Universe of our thoughts, feelings and memories is going to disappear. But what if our knowledge of ourselves is as incomplete as our previous knowledge about the Universe? The only thing we can do at this point is to draw upon analogies from the history of science which may prove to be instructive.

Prior to the advent of Einstein's theory of gravity, space and matter were considered as two fundamentally different entities. Space was thought to be a kind of three-dimensional coordinate grid which, when supplemented by clocks, could be used to describe the motion of matter. Space did not possess any intrinsic degrees of freedom, and it played a secondary, subservient role as a tool for the description of the truly substantial material world.

The general theory of relativity brought with it a decisive change in this point of view. Space and matter were found to be interdependent, and there was no longer any question of which was the more fundamental of the two. Space was also found to have its own inherent degrees of freedom, gravitational waves, which are associated with perturbations of the metric. Thus, space can exist and change with time in the absence of electrons, protons, photons, etc.; in other words, in the absence of anything that had previously been subsumed by the term matter. (Note that because of the weakness with which they interact, gravitational waves are exceedingly difficult to detect experimentally, an as-yet unsolved problem.)

A more recent trend, finally, has been toward a unified geometric theory of all fundamental interactions, including gravitation. Prior to the end of the 1970's, such a program, a dream of Einstein's, seemed unrealizable; rigorous theorems were proven on the impossibility of unifying spatial symmetries with the internal symmetries of elementary particle theory. Fortunately, these theorems were sidestepped after the discovery of supersymmetric theories. With the help of supergravity and superstring theories, one may hope to construct a theory in which all matter fields will be interpreted in terms of the geometric properties of some multidimensional superspace. Space would then cease to be simply a requisite mathematical adjunct for the description of the real world, and would instead take on increasingly greater independent significance, gradually encompassing all the material particles under the guise of its own intrinsic degrees of freedom.

Many people believe that consciousness, like space before the invention of general relativity, plays a secondary, subservient role, being just a function of matter and a tool for the description of the truly existing material world. Indeed, it is very easy to forget that our perceptions are the only "objects" which we really know to exist. The notion of matter appears only at the second stage of our investigation of the world, when we find out that it is convenient to describe the evolution of our perceptions using the concept of matter obeying laws of physics. These laws appear to be so universal, that at some stage it becomes very convenient to switch our position and to treat them rather than our perceptions as elements of reality. Therefore it is quite possible that nothing similar to the modification of the concept of space will occur with the concept of consciousness in the coming decades. But what if consciousness, like space, has its own intrinsic degrees of freedom, and that neglecting these will lead to a description of the Universe that is fundamentally incomplete? What if different perceptions should be considered as *really existing objects* What if it will turn out, with the further development of science, that the study of the Universe and the study of life should be inseparably linked, and that ultimate progress in the one will be impossible without progress in the other? Is it possible at all that eventually the theory of consciousness will encompass the theory of the "real world"?

All of these questions might seem somewhat naive. A healthy conservatism of science should keep us in the realm of the traditional approach to our world and to the laws of physics describing it, as long as this approach continues to be productive. However, it is impossible to avoid thinking about these questions in the context of quantum cosmology. For example, those who study quantum cosmology know that the wave function of the Universe, which determines the probability to find the Universe in a given state, *does not depend on time*. However, we know that the world around us changes. The resolution of the paradox is that we never ask questions about the whole Universe. We are dividing the Universe into two systems: an observer and the rest of the Universe. Then it can be shown that the wave function *of the rest of the Universe* does depend on time as measured by the observer, even though the wave function of the whole system is time-independent. In other words, *we see* the Universe evolving in time, even though the notion of change does not apply to the Universe as a whole. Does this mean that the Universe without us is dead, and only observers make it alive?

We do not have final answers to all these questions. Fortunately, in most situations of our daily life we can safely ignore them. However, working in the field of quantum cosmology without even trying to discuss these questions gradually becomes as difficult as working on the Big Bang theory without knowing why the Universe is so big and so homogeneous, why nobody has ever seen parallel lines intersect, why space-time is four-dimensional, and so on. Now, with plausible answers to these questions given by inflationary cosmology, one can only be surprised that prior to the 1980's, it was sometimes taken to be bad form even to discuss them. The reason is really very simple: by asking such questions, one confesses one's own ignorance of the simplest facts of daily life, and moreover encroaches upon a realm which may seem not to belong to the world of positive knowledge. It is much easier to convince oneself that such questions do not exist, that they are somehow not legitimate, or that someone answered them long ago.

It would probably be best then not to repeat old mistakes, but instead to forthrightly acknowledge that the problem of consciousness and the related problem of human life and death are not only unsolved, but at a fundamental level they are virtually completely unexamined. It is tempting to seek connections and analogies of some kind, even if they are shallow and superficial ones at first, in studying one more great problem, that of the birth, life, and death of the Universe. It may conceivably become clear at some future time that these two problems are not so disparate as they might seem.

This book makes several important steps in this direction. It contains many interesting insights and analogies. It also contains a discussion of some ideas which I might take in a different direction. But that is what makes this subject so exciting. The new cosmological theories are so unusual that sometimes we do not even know how to formulate correct questions, so there is no surprise that we do not always agree about the answers. We are trying to learn more about the Universe and about our own place in it. This is a challenging task, and we should try to fulfill it in a way which would match the efforts of those who were thinking about the fate of the Universe thousands of years before we were born.

PREFACE

Why worry about the fate of the universe? Aren't there enough preoccupations in life—taxes, mortgage payments, school, work, kids, parents, etc.—to fill one's days and nights with sufficient anguish? And how about those who are struggling hard just to put food on the table? Why should they trouble themselves with the ultimate cosmic questions?

As a cosmologist, studying the properties of the universe as a whole, I often find it hard to justify our spending billions of dollars for astrophysical projects such as space telescopes, when there are so many social problems yet to be solved. Yet the alternative—to retreat from our commitment to understanding the cosmos—would surely be a tragedy.

Our species aspires to push ever onward toward new frontiers. In this vein, the universe itself represents the ultimate challenge. Humankind will scarcely be content until it has conquered the mysteries of space and time, and has learned from where it has come and to where it is going.

This book ponders one possible answer to these riddles. Perhaps space periodically recreates itself like a snake shedding its skin (or, to employ the title image of this book, like a serpent continuously devouring itself). This is an ancient description of the cosmos that has been revived by contemporary theoretical physicists. Because there is no beginning nor end, it is truly a self-contained picture. Therefore, it avoids all the conundrums associated with the question: what came before the start of time?

There are other competing models that suggest alternative cosmological approaches. In some, the universe has a starting point but no stopping point; in others, time has rigid boundaries. To resolve this debate, we need to peer farther and farther into space, with instruments such as the Hubble Space Telescope.

Space science comes with a cost, and requires a firm societal commitment. I believe that this generally is money and effort well spent. Studying the cosmos involves replacing, for a time, one's personal considerations with universal matters. I believe that it is well worth the exchange.

So let's sit back, relax, put aside our ordinary concerns for the time being, and contemplate the ultimate destiny of the universe.

CONTENTS

INTRODUCTION   1

PART 1 THE TRADITION OF CYCLES   9

CHAPTER 1   THE ENDLESS DANCE OF SHIVA   11

    Invitation to the Dance   11
    Cult of the Serpent   14
    Great Years   19
    Harmony of the Spheres   25
    Wheels of Destiny   34
    Nothing New Under the Sun   40

CHAPTER 2   ETERNAL RETURN   45

    Nietzsche's Encounter with Eternity   45
    The Atomism Debate   50
    Time's Arrow   57

Contents

       Heat Death   62
       The Recurrence Controversy   64
       Eternal Return Revisited   69

PART 2 THE MODERN VIEW OF THE COSMOS   73

CHAPTER 3   THE EXPANDING HEAVENS   75

       Unholy Designs   75
       Skygazers   82
       Cosmic Evidence   87
       The Background Hiss of the Great Serpent   96

CHAPTER 4   FULL OF SOUND AND FURY   105

       An "Ourobouric" Quandary   105
       The First Few Seconds   109
       The Uniform Sky   113
       Crossing Horizons   117
       The Age of Inflation   119
       Galactic Seeds   127

PART 3 THE QUEST FOR OUR UNIVERSAL DESTINY   133

CHAPTER 5   MAPPING OUR FATE   135

       Singular Genius   135
       From Mass to Motion   139
       The Power of Omega   144
       The Price of Inflation   151
       The Big Crunch   154

Contents

CHAPTER 6 THE SHAPE OF CREATION 157

Longevity Tests 157
Getting in Shape 159
The View from Flatland 165
The Fifth Dimension 167
Beyond the Fifth Dimension 169
The Cosmic Deficit 171

CHAPTER 7 GALACTIC SPEEDING TICKETS 177

The Robot Pitcher 177
Ladder to the Stars 179
The Great Debate 183
Supernova Speedometers 187
The Magic of MERLIN 189
Hubble Trouble 192

CHAPTER 8 THE SEARCH FOR MISSING MATTER 199

Weighing the Universe 199
Hidden Planets 201
Stellar Carousels 204
MACHO Men 207
Intergalactic Shadows 211
Running Hot and Cold 215
The Closing Circle 217

Contents

PART 4 UNRAVELING THE CYCLICAL SERPENT 219

CHAPTER 9 REVERSE PERFORMANCE 221

The Sky Is Falling 221
Confronting Collapse 222
The Currents of Time 227
Playing It Backward 232
Time Without Boundary 236
The World That Devoured Itself 244

CHAPTER 10 AFTER THE CRUNCH 247

The Rhythm of Eternity 247
Beyond the Omega Point 248
The Entropy Crisis 252
Reprocessing Space 255
Only the Names Are Changed 258
Surviving the Crunch 261
Everlasting Life 263

CHAPTER 11 OTHER COSMOS 267

Bubble Universes 267
Eternal Inflation 270
The Best of All Possible Worlds 273
Natural Selection 275
The Meaning of It All 277

EPILOGUE 279

REFERENCES 281

RELATED READING 283

INDEX 293

INTRODUCTION
THE FARTHEST SUPERNOVA

> The World's great age begins anew,
> The golden years return,
> The Earth doth like a snake renew
> Her winter weeds outworn
> -Shelley, "Hellas"

Nothing is more horrifying than the thought of the end of the world, of the literal possibility that everything we know and have ever known is doomed to perish, never again to return. As in the case of personal death, universal demise offers absolutely no hope for escape that might grant some comfort. But, to make matters worse, unlike individual extinction, world destruction provides no solace in the survival of next of kin. Obviously, it's no wonder that we hope the world will never end, that time will go on forever.

In July 1054, Chinese astronomers of the Sung dynasty recorded a cataclysmic event in their skies, one that presumably many believed was a signal of the end of the world itself, or perhaps the beginning of a new cycle of time: the sudden annihilation of a distant star or, in modern terms, a supernova explosion. As seen at the K'ai-feng observatory, a stellar light source began to grow in intensity, expanding over days, its yellow glow gradually dwarfing many of the other prominent features of the night sky and becoming visible in daylight. Then, three weeks later, the eruption started to subside – restoring, over more than a year's time, the sky's appearance to normal. It was not until hundreds of years later, after the invention of the telescope, that astronomers could observe the star's colorful, crustacean-shaped, hazy remnant, which they named the Crab Nebula.

Today we view distant supernovas as sources of fascination, rather than objects of fear. Without a doubt, the explosion of a star is a wondrous sight to behold—as long as the burst is comfortably located many trillions of miles away. Telescopic images of the results of supernova eruptions have proven to be among the most beautiful in all of astronomy. Pictures of the Crab Nebula and related supernova remnants have decorated countless science classrooms and laboratories. Only if supernovas were nearby would we have to worry about our own extinction. The blasts of fire generated in their stellar cataclysms would be enough not only to destroy the exploding star's own planetary system, but also to disrupt the motions of other stars and planets, such as ours, in the immediate vicinity. If our own sun (assuming it is massive enough, which it probably isn't), or a sufficiently close star, was destined to undergo such a fate, and we even received ample warning, there would be little for us to do but flee, if we possibly could.

However devastating they would be to nearby planets, it is wrong to conclude that supernovas are wholly destructive. Quite the contrary; if we look at the larger picture, supernovas might represent kinds of cosmic recycling devices. It turns out, in fact, that the regenerative tasks needed for the production of most of the vital elements found on Earth took place in the fiery cauldrons of primitive stars; these materials were later released into space when the stars exploded. Thus, oxygen, nitrogen, calcium, iron and carbon—the essential ingredients of life and more—came to us by way of supernova bursts, propelled through the interstellar void and formed into the very fabric of our planet and its atmosphere. Supernovas, then, rather than representing astral eradicators, might be viewed instead as celestial resurrectors, generating new matter from the relics of the old, and planetary life from stellar death.

Perhaps it is not surprising that regeneration occurs so often among the stars, since it is so commonplace here on Earth. Cyclical features abound in nature. Over the millennia, dozens of new islands, teeming with life, have formed out of the settled ash and dried lava of regions once destroyed by sudden volcanic eruptions. Furthermore, it is well known that during ruinous African droughts, the rich soil, swept up in dust storms, blows across the Atlantic to cause renewed growth in the rainy Amazon basin. And, an example of natural recycling comes from the distant past, the demise of the dinosaurs, caused, perhaps, by climate changes resulting from a large meteor or comet hitting the Earth (as many scientists theorize). By wiping out the giant reptiles, this terrible impact provided a tremendous boon for new species of mammals – including, eventually, the human race.

Clearly, a certain measure of natural harmony, of balance between creative and destructive elements, maintains itself on Earth. This is exemplified not only in the rise and fall of geological formations, but also in the rhythmic behavior of plant and animal life associated with the procession of the seasons. In autumn, the leaves decay into topsoil and provide the raw materials for spring's renewed growth. The melting torrents of winter snow provide the fresh water needed for trout to flourish in the summer. For all of nature's creatures, then, for all intents and purposes, terrestrial history is endlessly renewable.

What then of the cosmos itself? Might the entire universe, at some point, undergo total renewal? Might "seasons of growth and decay" form a delicate balance in the universe at large, as well as here on Earth? And if indeed there are cosmic, as well as terrestrial, epochs of creation and annihilation, then what would become of the human species (or any form of intelligent life) if the universe were to recycle itself completely? Need we continue to fear the idea of time's end, or might we find comfort instead in the notion of endless temporal cycles?

These fascinating questions – our subject of exploration here – have become increasingly the focus of theoretical and experimental astrophysics. In recent years, scientists have produced a detailed picture of the origins of the cosmos, and have confirmed it through careful experimentation. Because this model of the beginning of the universe is so well developed, many researchers believe that we will soon discover the fate of the universe as well.

According to standard cosmology, the observed physical universe was produced billions of years ago in a catastrophic outburst, christened "the Big Bang" by Fred Hoyle. Evidence in support of this model appears all around us in the form of microwave radiation relics of the explosion, as well as in the visible recession (moving away from our own galaxy) of all other galaxies, sparing those in a small local cluster. These indications lead most scientists to believe that the cosmos is presently expanding, as if all of its constituent galaxies were situated on an increasingly inflating balloon.

What is not yet known is whether or not the balloon will burst. Will the cosmos continue to expand forever, or will it someday stop its dilation and begin to collapse in on itself, in a process often called "the Big Crunch?" Neither cosmological theory, nor astronomical measurement have yet revealed the truth. Here, we'll mainly explore the Big Crunch scenario (my own philosophical preference), keeping in mind that the verdict is not yet in.

Assuming that the universe were eventually to recontract down to a single point, its subsequent fate would be most uncertain. Some theorists have argued that this crushing moment would represent the end of time itself. Others have asserted that a "Big Bounce" would then occur, propelling the cosmos back into a second Big Bang explosion. Then, according to this model, the new burst would be followed by further periods of expansion and collapse, with universal destruction and regeneration taking place again and again. Whether or not this sequence would eventually come to a halt, and whether or not reality would become radically altered after each bounce, are the subjects of additional debate.

One might think that these questions would be impossible to answer; after all, they involve hypothetical events that would take place billions of years in the future. Surprisingly, though, many astronomers now believe that they are coming close to resolving some of the issues that impinge on the ultimate fate of the cosmos. By carefully measuring the distances and recession velocities of distant celestial objects, they soon hope to determine whether or not cosmic expansion will eventually grind to a halt and reverse itself.

Introduction

In April 1992 a team of British and American scientists, led by Saul Perlmutter, Carl Pennypacker, and Gerson Goldhaber of Berkeley, while scanning the skies over the Canary Islands with the Isaac Newton telescope, discovered an extremely distant supernova. They soon realized what a boon the observation of this stellar blast would be to our understanding of the Big Bang explosion. After several months of detailed analysis of this event, they announced that they had determined that the supernova had exploded some five billion years ago in a galaxy over 25 billion trillion miles away; it was the oldest and farthest supernova ever charted. The significance of this finding was that it allowed scientists, for the first time, to measure the deceleration rates of some of the most distant points in the cosmos, thus enabling them to calculate with far greater accuracy than before the amount that the universe is slowing down in its expansion.

The reason supernovas are so useful for this purpose is that they represent what is called "standard candles": celestial objects with known average brightness that can be used in assessing distances; the more remote they are, the smaller amount of their light reaches Earth and the dimmer they appear. Thus by comparing the apparent intensity of a newly discovered supernova to the known value of its brilliance, one can obtain a direct indication of how far away it is. Finally, by combining these distance results with determinations of the velocity of the galaxy containing the supernova, one can roughly estimate the expansion rate, for all times, of the universe as a whole.

The Berkeley researchers are now preparing for a massive survey of distant supernovas. This will enable them to hone their results, which so far neither support, nor exclude, the possibility of cosmic collapse. In order to achieve the goal of resolving the universe's fate, they are using computers to track more than 2500 different galaxies simultaneously. Other groups are similarly working on collecting numerical data for a further assessment of this situation. And the use of supernovas as cosmological tools is but one of the many methods astronomers and physicists are now employing to try to determine, once and for all, if the universe will undergo limitless cycles of expansion and contraction.

The idea of a recurrent cosmos isn't a new one, of course. Throughout history, stargazers who recorded unusual astronomical events, such as supernovas and comets, presented their societies with dire warnings of imminent apocalypse. The thought that everything they knew and cherished might come to an abrupt halt was simply too much for them to bear (as it is for many of us today). Hence, they refused to entertain the notion that the world could ever really end. Faced with the terrors associated with the possible total annihilation of the universe, most premodern cultures thus chose to believe in a cyclical time model of a cosmos undergoing infinite recurrence. They saw the world as an interaction of equally powerful constructive and destructive forces, as a cosmic drama that would continue forever.

This belief in eternal cosmic repetition is reflected in the myths and symbols of the ancient world. From the holy shrines of ancient India to the great temples of classical Greece, philosophers and storytellers rendered epic tales of battles between the forces of good and evil. As depicted, these struggles would last for eons. Though one side would prevail for a certain period, neither would ever win decisively.

Introduction

To appreciate fully the research and hypotheses of contemporary scientists on the cutting edge of astronomy, it is instructive to examine how the ancients perceived the cosmos. Their vision of a universe in cycles, as conveyed in their art, music and dance, prefigured many of the controversies that continue to rage in the science of cosmology.

PART 1

THE TRADITION OF CYCLES

THE ENDLESS DANCE OF SHIVA

I have known the dreadful dissolution of the universe. I have seen all perish at the end of every cycle. Ah, who will count the universes that have passed away or the creations that have risen afresh, again and again, from the formless abyss of the vast waters?
-Brahmavaivarta Purana (Hindu sacred text)

Invitation to the Dance

In a darkened theater on a bank of the sacred Ganges river, the time-honored ceremony begins. A robust, bare-chested young man wearing a girdle of serpent-skins, earrings of serpent-tails, and a cord of serpents, sits passively, cross-legged on the stage in a state of deep contemplation. Nearby, a slender young woman, draped in a long, colorful *saree*, dances around him in a joyful, energetic ballet.

Suddenly, a horrifying figure, a demon in the shape of an elephant, leaps forward onto the stage and attempts to abduct the woman. While the audience gasps, the young serpent-man jumps up, grabs a spear, and runs after the adversary. He hurls imitation bolts of lightning at the would-be abductor, and dances around him in a death march until the defeated creature staggers and falls to the floor.

Then, the serpent-clad victor executes yet another dance: a celebration of the endless creation and destruction of the world. Slowly and solemnly, as if to reflect his complete indifference toward his predestined triumph, he repeatedly waves his arms and legs in a graceful, sinuous pattern. Finally, the battle won and the pageant concluded, he sits down and resumes his cross-legged position of peaceful meditation.

The Dance of Shiva ceremony, complete with actors portraying serpent-god Shiva, his beautiful and divine consort Shakti, and his hideous, demonic archenemy (usually, but not always, in the shape of an elephant), is one of the oldest religious rituals in the world. It has been performed for thousands of years, by thousands of actors, on thousands of stages. Awesome and inspiring to behold, it is a source of tremendous devotion for millions of Hindus. Often, this ritual is accompanied by considerable offerings to Shiva of food, flowers, and other gifts. Devotees have been even known to practice some of the most excruciating forms of self-immolation while overwhelmed by the passion of the rite.

Shiva (the Destroyer), the source of all this tribute, is, by far, the oldest of the gods forming the Hindu sacred triad, predating Brahma (the Creator) and Vishnu (the Preserver). Shiva was worshipped in ancient India even before the Vedic scriptures, the basis of Hinduism, were written. In these earlier notions, predating the coming of the Aryans to India, Shiva was seen as an autonomous figure—a personal deity with both creative and destructive qualities. Only later, as these tribal customs were incorporated into the Hindu code of belief, did Shiva come to be regarded as but one aspect of a multifaceted, all-encompassing holy trinity.

Truly, to understand Hindu cosmology, one must fully appreciate the extraordinary richness of the Dance of Shiva. The Indian concept of time is cyclical—epochs of universal creation and destruction succeeding each other an indefinite number of times. Shiva's dance, which is the explicit manifestation of this belief in cycles, is a tangible expression of the ceaseless rhythm of the cosmos. As Shiva gestures gracefully, but resolutely, his swaying arms and legs symbolize the bringing about and subsequent annihilation of every atom of matter and energy in the universe. His austere, balanced motions reflect his complete impartiality toward this cyclic succession of emergence and dissolution; he favors neither birth nor death, maintaining instead a measured equilibrium between creative and destructive principles. All in all, his dance is a celebration of the fleeting nature of physical reality. As Fritjof Capra, author of *The Tao of Physics*, puts this: "The Dance of Shiva is the dancing universe; the ceaseless flow of energy going through an infinite variety of patterns that melt into one another."[1]

In the 10th and 12th centuries, Indian sculptors created masterful stone and bronze renditions of the Dance of Shiva. Shiva was represented in these depictions as having four arms and two legs, each oriented in a different direction and engaged in a different activity of allegorical significance. In stark pictorial contrast, the upper right hand carried a drum, representing the rhythmic process of creation, and the upper left hand held a flame, symbolizing the fiery power of destruction. The positions of two other limbs signified the protective aspects of the deity; while the right leg was shown stepping on a demon, the lower right arm was portrayed in a comforting gesture indicating that there was nothing to fear. Finally, the two remaining limbs, lifted high up in the air, provided a vibrant visual depiction of the lofty emancipation from fear and ignorance that one might obtain by worshipping Shiva.

The idea of freeing oneself from illusion by gaining knowledge about the true nature and purpose of the universe is an all-important concept in Hinduism. In Sanskrit, the word for cosmic illusion is *maya*, and it is release from *maya* that is a primary goal of all devout Brahmans. Hindu thought insists that we should not believe our senses, which tend to indicate that the world is solid and permanent, but instead should trust our faith, which leads us to realize that everything around us is ephemeral.

Thus, the colorful Dance of Shiva ceremony and the beautiful Shiva sculptures remind us that the universe is ever-changing, its matter and energy in a constant state of flux. The wildly twisting form of this ancient Indian deity is therefore a perfect metaphor for the regenerative transformations that continually alter the very fabric of the cosmos.

Cult of the Serpent

In Indian mythology, Shiva is often addressed by the epithet "King of Serpents." Indeed, in all of his traditional depictions, he is seen either holding serpents, crowned by serpents, or draped with serpents—and sometimes all three. Clearly, his association with these venomous creatures serves to enhance his perceived powers and grants him an aura of invulnerability.

Serpent worship is one of the most common forms of religious ritual. Not only in India, but around the world, serpents have been venerated for generations. From the

Aztecs of Mexico, who prayed for the return of the legendary white-bearded serpent-god Quetzalcoatl, to the Dahomeans of West Africa, who used coiled snakes as symbols of their divinity Da, the reptilian motif is ubiquitous. Snake patterns are similarly omnipresent in Australian aboriginal art, often representing thunder, lightning, and rain spirits. In China, the serpent and dragon are usually interchangeable and are typically seen as harbingers of malevolent power. Serpent-shaped mounds that are the remains of Native American burial sites—can be found in central Ohio. Every Bible school pupil is, of course, familiar with the story of Adam, Eve, and the serpent, a tale reflecting, perhaps, the ancient Hebrews' suspicion of their neighbors' veneration of snakes. Indeed, archeological digs all over the Middle East have turned up numerous shards of pottery featuring intricate serpentine patterns. The Babylonian Ishtar, great goddess of periodic cosmic rebirth, is always portrayed with a snake scimitar in her left hand. And the menacing serpent Sata, said to be seven cubits long, prominently appears in the Egyptian Book of the Dead as a symbol of daily rejuvenation. "I am the serpent Sata," it is written, "I die and am born again each day."

It is precisely these mysterious powers of "reincarnation" that makes the snake such a poignant symbol of cyclical time. Snakes, by shedding their skins during their annual season of hibernation, seem to prolong their lives, revitalize themselves, and even gain in size and strength. Thus it is natural to consider the regeneration of serpents as emblematic of the periodic renewal of the universe.

Other qualities possessed by snakes similarly account for their close symbolic association with cosmic cycles. Their undulating, wavelike movement suggests emergence of the vital elements from the ocean of universal possibilities and reabsorption of these substances back into the same dark waters. Also, their staring, lidless eyes seem to convey deep knowledge about the mysteries of the world, including secrets of death and rebirth to which humans are not privy. And while their phallic shapes evoke masculine images of potency, their egg-laying powers imply feminine aspects of fertility, leading to an overall enigmatic impression of androgynous self-reproduction.

Perhaps, though, the property of serpents most evocative of ceaseless universal cycles is the ability of these creatures to coil. Their flexible spines enable them to form complete circles—tail-to-mouth—which humans have interpreted as self-ingestion, symbolizing the periodic coming forth and disappearing of the cosmos.

It is not surprising, then, that the coiled, self-devouring serpent, usually called *ourobouros* (Greek for "tail-consuming"), is a frequent mythological symbol of cyclic time. And, while it is improbable that any snake has ever even attempted to swallow itself, the *ourobouros* motif has been an emblem of cosmic reincarnation for at least six thousand years.

One of the earliest known examples of the use of *ourobouros* dates back to the Yang-shao culture (about 4500 B.C.) of China. Vases have been found from that period that include multihued, self-devouring serpent designs, differentiated into darker and lighter cross-hatched halves. These patterns clearly suggest cyclic periodicity, as in the manner of day following night or as in the case of the phases of the moon. It is likely that the familiar yin–yang symbol—the circular, black and white, Taoist insignia—has evolved from these more fundamental pictures.

Other instances of cyclical serpent imagery include five-thousand-year-old relics of the Elamites, found in exca-

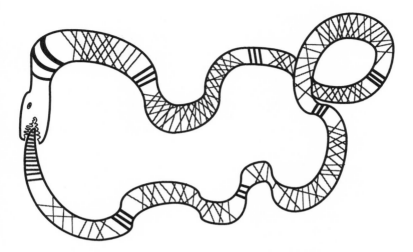

Figure 1. Ourobouros, the cyclical serpent.

vations at Susa (Iran), their capital, as well as sacred artifacts of the Dahomean people, dating back to a much later period. In many of these myths, the serpent encircles the whole world, and represents the circuitous flow of the waters surrounding the Earth. The Jainas, a non-Hindu religious community in India, use a variation of *ourobouros* to express contrasting aspects of repetitive time. They contrast *utsarpini* (literally, "up-serpent"), the ascending epoch of rising hope, with *avasarpini* ("down-serpent"), the descending epoch of imminent annihilation, in a design symbolizing the eternal cosmic circle. And yet another version of this symbol, used by the Arabian alchemists, is a prototype of the modern symbol for infinity.

Finally, in perhaps its most famous appearance, the *ourobouros* emblem was widely employed by the ancient

Egyptians to denote the transmigration of souls and the coming of new world epochs. Horapollo, or Horus Apollo, the author of a famous treatise on Egyptian hieroglyphics, relates that:

> When the Egyptians would represent the universe, they delineate a serpent bespeckled with variegated scales, devouring its own tail, the scales intimating the stars in the universe. The animal is extremely heavy, as is the Earth, and extremely slippery, like the water. Moreover, every year it puts off its old age with its skin and becomes renovated like the universe. And the making use of its own body for food implies that all things whatsoever, which are generated by divine providence in the world, undergo a corruption into them again.[2]

So, for the Egyptians the image of a serpent forming a ring by swallowing its own tail is expressive of the notion that the heavenly firmament is continually producing and reabsorbing both material objects and spiritual beings. Souls, they believe, could never be created nor destroyed but could only pass from one body to another in an endless chain of reincarnation. This idea of cyclical reawakening pervades all aspects of their society and affects even their architecture. Certainly, the construction of the great pyramids, stocked with the mummified remains and the worldly possessions of the Egyptian rulers, is testimony to the belief that the spirits of these dead kings would someday return and reclaim these bodies and provisions. Nothing is ever gained or lost, the Egyptians would seem to assert; the abacus of history must register additions and subtractions in equal amounts.

Egyptian philosophy further emphasizes a deep personal identification with the cycles of nature. The annual recessions and floods of the Nile river, source of sustenance for all of Egyptian society, are seen as indicative of mysterious restorative properties. It is for this reason that submergence into the Nile is equated with holy redemption and revitalization (and another reason why serpents, appearing and disappearing into these waters, are venerated). Prayers recited during mummification, for instance, refer to this watery domain as the resting place of souls on their lengthy posthumous journeys. Invoking Osiris, god of cyclic death and renewal, these entreaties celebrate the eternal rhythms of nature as ultimate manifestations of divine providence. And, therefore, it comes as no surprise that in Egyptian cosmology Osiris is often represented by the ultimate symbol of natural periodicity: *ourobouros*, the cyclical serpent.

Great Years

When the Egyptians would discuss their temporal philosophies with the inhabitants of neighboring regions they would find much agreement. All of the ancient peoples (with one notable exception—the Hebrews, from whom stems contemporary Western notions of linear time) believed in the concept of cycles. In contrast to generally held modern-day beliefs, history, for them, was not a uniform stream of destiny, steadily flowing forward, bringing one to new locales, but, instead, a rapidly spinning whirlpool from which there was no escape. In their view, the currents of circumstance churn at first to the left and then to the right, to and fro, hither and thither, but ultimately bring one back to one's starting place.

Considering these cultures' total dependence on the periodic rhythms of nature, it isn't hard to understand their supreme faith in endless recurrence. The motions of the sun, the stars, the wind and the tides, which determined the exact times of leveling and plowing, of sowing and reaping, of drought and deluge, and of feast and famine, were of crucial importance to these agriculture-based societies. And those who gained mastery of these natural timetables tried to achieve prosperity by tailoring their behavior to celestial patterns.

Holidays, in particular, were seen as good times to synchronize one's own pulse with the heartbeat of nature. By involving oneself in ritual activities that celebrate seasonal abundance, such as spring carnivals, harvest festivals and New Year's revelry, it was thought that one might somehow achieve wealth, happiness and eternal youthfulness. Moreover, because human fertility and agricultural fertility were thought to be closely related, weddings were planned to coincide with these annual holidays.

It was thus deemed important to determine the correct dates and times for the commencement and completion of these seasonal rituals. Therefore, in many cultures stargazers were employed for the task of ascertaining planetary and stellar motions, as well as for predicting eclipses and conjunctions (planetary alignments). The results of these astronomical studies were often used to devise calendars.

Consider, for example, the other pyramid-building culture of the ancient world, the Mayans of Central America. This highly advanced pre-Columbian civilization developed a 260-day ritual calendar based on their sacred year, the *tzolkin* (they considered 260 to be a magical number). Because the Mayan astronomers also recognized the importance of the solar year of 365 days, they further developed the idea of the so-called Calendar Round: a great cycle of time, lasting 52 years (or 18,980 days, which is the least common multiple of 260 and 365), during which the solar cycle and *tzolkin* would come back into realignment. Thus, after one Calendar Round, the New Year's day of the *tzolkin*

Figure 2. Aztec calendar stone. (Neg. No. 334432, Courtesy Department of Library Services, American Museum of Natural History.)

would once more coincide with the New Year's day of the solar year.

When the Aztecs later adopted the Mayan calendar, they began to attach a ritual significance to this 52-year time span, and grew to embrace the superstitious belief that the end of the world would come at the close of one of these cycles. Therefore, during the last days of these intervals,

Aztec zealots would abandon their homes and flee to the hills in fear of an all-consuming earthquake or other natural catastrophe. Finally, after such a period had passed, they would joyfully return to their cities and offer much homage to the gods in gratefulness.

Many other calendar-based early civilizations similarly assigned catastrophic properties to the completions of astronomically based, multiple-year time cycles, often called Great Years. The ancient Chinese, believers, as discussed, in a cyclical time represented by both a self-devouring serpent design and the famous yin–yang symbol (a circle equally divided into curved black and white segments), also attributed creative and destructive characteristics to Great Years. For them, the 12-year interval called a Yin, the time during which Jupiter nearly completes its sidereal period (returns to the same position in the sky), was an epoch of enormous significance. Each of these eras constituted the cyclic succession of great harvests, famines, floods, and droughts. Moreover, a "world cycle" period of 23,639,040 years, during which the then-known celestial bodies would execute a complete repetition of their combined motions, was considered to be the lifetime of the universe. According to the ancient Chinese, at the end of this supreme interval, a global cataclysm would occur and the world would then begin anew.

The Chinese were not alone in their belief in cosmic cycles on such a grand scale. Far across Asia, in the fertile river valleys of Mesopotamia, Babylonian soothsayers were similarly forecasting universal catastrophe, based on their own detailed measurements. And, due to Babylonian society's virtually unlimited deference to the prognostications of conjurers, these predictions carried enormous weight

Endless Dance of Shiva

Figure 3. Yin–Yang, symbol of eternal cycles.

there and wielded considerable influence over day-to-day life.

Most historians agree that Babylonia is the birthplace of astrology. From time immemorial, in Babylonia and other early Mesopotamian civilizations, divination was used to determine the fates of kings and common folk alike. Methods for forecasting had included casting lots, observing animal behavior (the proclivities of cats, for instance), and reading omens obtained from ritual incantations. So when reliable procedures for plotting the paths of the stars and planets were developed during the Chaldean dynasty (625 to 539 B.C.), these techniques naturally were incorporated into the Babylonian predictive method and astrology was born.

Babylonian priest-astrologers looking for divine portents were particularly interested in unusual celestial phenomena, such as eclipses and planetary conjunctions, rather

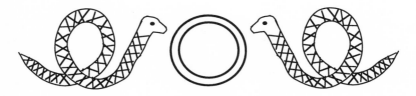

Figure 4. A Chinese motif: Two serpents facing a ring.

than more typical astronomical behavior. Once detected, they would then associate these heavenly signs both with impending natural catastrophes, such as invasions of locusts or devastations of crops, and with imminent political upheavals, such as overthrows of kings or the slaying of armies. For them, the configurations of the planets and the patterns of human lives mirrored each other.

Much of what we know about this period comes down to us from the writings of Seneca, a Roman statesman and philosopher who lived about the time of Jesus. Seneca was particularly interested in the philosophy of Berossus, a Babylonian priest who immigrated to the Greek island of Cos around 290 B.C. Although fragments of Berossus' original writings have been unearthed and translated, it is particularly through Seneca's perceptive interpretation that we gain insight into the remarkable realm of Berossian thought.

Berossus, according to this detailed depiction of his ideas, was a keen observer of the motions of the celestial bodies and, like many of his predecessors, had assigned allegorical significance to their shifting configurations. Embracing the notion of a strict correlation between human and cosmic developments, he developed his own chronology of world events, dividing the whole of history into four distinct phases.

The first era, lasting 1,680,000 years, began with creation and lasted until the birth of the first of the "ancient kings" of Mesopotamia. Following this was the second era, lasting 432,000 years, comprising the age of the 10 "ancient kings." This era ended with a great world flood, a period corresponding to a special conjunction of the planets in the zodiac region of Capricorn, in which they were precisely arranged in a straight line.

After the deluge came the third world era, which lasted exactly 36,000 years and ended with the death of Alexander the Great. The final era, including the time of Berossus, would last only 12,000 years. Immediately following this interval, the planets would realign themselves again—this time in Cancer—and there would be a great fire that would completely destroy the surface of the Earth.

In Berossus' view, these four phases would repeat themselves again and again, as the planets returned repeatedly to their special alignments in a Great Year lasting over two million years. Thus, history, as he saw it, was entirely cyclic, dependent, as he believed, on the precisely periodic (like clockwork) orbits of the heavenly spheres.

Harmony of the Spheres

Seneca's reflections on Berossus appeared in the first century A.D., well past the prime of Greco-Roman science, during a period of general decline. Therefore, it is not surprising that the Roman critic read into the Babylonian philosopher's writings a sense of utter futility about the meaning of human achievements in the face of eternal recurrence of history. Seneca's feelings about the growth of vice and decline of learning in Rome lent an aura of despondency to his commentary, exemplified in his predictions about the end of the human race:

> A single day will see the burial of all mankind. All that the long forbearance of fortune has produced, all that has been reared to eminence, all that is famous and all that is beautiful, great thrones, great nations—all will descend into the one abyss, will be overthrown in one hour.[3]

Other Roman historians of the age, from Cicero to Pliny the Elder, similarly believed in the doctrine of the Great Year and the decline and fall of humankind in a worldwide catastrophe, followed by an epoch of global regeneration. And in a similar scolding manner, they linked the inevitable destruction of the Earth to the widespread decay around them of culture and values. Cicero, in particular, warned against harboring any vain illusions of fame or fortune, for all would be destroyed in the end.

But in spite of Seneca's allusions to Berossus, it's clear that the Roman doctrine of cycles derived primarily from the works of classical Greek philosophers, not the findings of Babylonian astrologers. For by the time Berossus emigrated to Cos, Greek society had already produced a number of important writings—dating several centuries back to Pythagoras—commenting on the prospects for an ever-repeating cosmos.

The word *cosmos* itself (in Greek, literally: beautiful harmony), in its current meaning of an all-embracing world system, was invented by Pythagoras, who founded his school of philosophy on another Greek island, Samos, around 531 B.C. The Pythagoreans sought an explanation for the periodic motions of the planets, and turned to mysticism as a means of resolving this question. In their mystic quest, they developed a detailed numerology and extolled the virtues of rational numbers, ultimately seeking to unravel the riddles of mathematics and thus master the secrets of the universe.

Indeed, for the Pythagoreans, numbers were seen as holy instruments; to appreciate their significance was a way of getting closer to the divine. The cyclical order of the heavens could only be explained, they thought, by extracting the numerical values from its components (orbital times, for instance), and determining the proper proportions between these quantities. This sounds all very scientific—quite the contrary. Although the Pythagorean school was on the right track towards modern cosmology in sensing a connection between astronomy and mathematics, the math they used had nothing to do with physical measurement at all. The numerical quantities with which they calculated were chosen more for their intrinsic beauty as numbers than for their utility in solving equations of motion.

Pythagoras and his followers were extraordinarily abstract—much like the Laputans in Swift's *Gulliver's Travels*: with their feet firmly planted in the clouds and their gaze permanently glazed over in abstruse reflection. Not a single experimenter among them; the Pythagoreans believed that it was far better to be a detached onlooker than an active participant. This disdain for the player is clearly demonstrated in their parable about the Olympic Games, which states that those who watched and recorded these games were of superior breed to those who took part in them as athletes. It is interesting to compare this dichotomy with the modern one in the world of scientific research between theorists and experimentalists; in fact, the word *theorist* derives from the Greek expression "to look on." The Pythagorean school, then, represented theory completely divorced from experiment, drawing of conclusions in isolation from empirical data.

Present-day theories of the universe, of course, stem from both detailed astronomical observations and complex mathematical relationships based, at least in part, on these measurements. Interestingly, Babylonian cosmology contained stargazing without much mathematics, while Pythagorean cosmology embraced numerics without much skywatching. Thus, because of their mystical obsessions—the Babylonians with unusual astral phenomena, the Pythagoreans with unusual numbers—neither group had the insight to strive for a complete and accurate model of the heavens.

The cyclical cosmic model that the school of Pythagoras proposed was based on two numerological principles: the mysterious powers of the number 10 and the harmonic properties of the musical scale. Ten, they felt, was a perfect number that encompassed all features of the universe. Somehow, then, the physical world was constructed from the first 10 digits, they surmised. To better understand this construction, they examined the characteristics of simple ratios of these digits; for example, 2:1, 3:2, and the like. Much to their amazement, they found that these same proportions are omnipresent in the world of music. Thus, they concluded that the planetary spheres lie in direct correspondence with the tonal harmonies, produced by the sounds of a celestial musical instrument as if it were plucked. This is the so-called harmony of the spheres.

Ironically, in spite of the seemingly *ad hoc* nature of the Pythagorean proposals, we now know that there is, in fact, some truth contained within them, albeit on a minute, rather than planetary, scale. Today physicists recognize in quantum mechanics a profound connection between the harmonies of music and the fluctuation states of elementary particles. For example, the probability waves of electrons

Endless Dance of Shiva

(indicating the most likely positions of these particles) surrounding a hydrogen nucleus demonstrate patterns of peaks and valleys similar to those of the vibrational harmonies of a plucked guitar string. Thus, remarkably, on the subatomic level the Pythagorean correspondence between the musical scale and the physical universe has at least some relevance to current modes of thinking.

Applied, though, to the planets, Pythagoras' model falls short. As we know today, the motions of the bodies that make up the solar system exhibit behavior manifestly unlike the vibrations of musical instruments. Furthermore, another of the predictions of Pythagoras and his followers—that because 10 is a perfect number, there must be 10 heavenly bodies—has similarly proven to be false (since only nine celestial objects were visible in their day, including the sun and moon, they postulated a "counter-Earth" to be the tenth). So, in terms of concrete achievements in the field of astronomy, little can be said for the school of Pythagoras.

From the age of Pythagoras until the time of Plato, little more was said by the classical Greeks about the nature of the universe. The philosophers during this period (roughly the fifth century B.C.) who did discuss cosmological questions, such as Heraclitus of Ephesus and Empedocles of Akragas, among others, were just about as mystical as the Pythagoreans and unfortunately even more vague. Heraclitus, for example, was known even to his contemporaries as "the Obscure One." He advocated the notion that all events in the cosmos, including those of human lives, will eventually repeat themselves, but also enigmatically insisted that "we will never step into the same river twice."

Although the ideas of Heraclitus about cosmic cycles were vague, he presented them in a confident, methodical manner. Empedocles, in contrast, wrote in a flowery, emotional style almost unknown to philosophers of his day. A poet and statesman by occupation, and incredibly vain, he promised his followers magical powers if they would obey his commands and worship him as a god. His far-fetched notions barely made a dent in the history of cosmological theory.

Unlike the cold, mathematical model proposed by the school of Pythagoras, Empedocles' cosmos is one brimming with passion, consisting of the interplay between two primary forces, Love and Strife. Love is the power that brings discordant elements together into a single, unified body; Strife is the agent that separates them again. The history of the world to Empedocles then, is a series of epochs in which Love dominates and the cosmos is harmonious, followed by epochs in which Strife prevails and the cosmos is discordant. And, in Empedocles' view, these universal cycles continued indefinitely, alternating between one and the other principle.

Although all three of these early Greek philosophers believed in cyclical time, there was considerable disagreement as to its impact on human identities. While Pythagoras argued for exact, numerical recurrence of individual human lives, Empedocles advocated a far less rigid theory of cycles, in which the human race as a group, but not individually, would experience endless rebirth. Heraclitus, by contrast, took the middle ground, and spoke for individual recurrence with a change of identity upon rebirth; that is, one would be born again without realizing it.

Throughout the history of the classical age, the prevailing attitude toward the question of exact, versus approximate, repetitions of human experience continued to shift from one position to the other and back again. Compare, for instance the time of Plato (428–348 B.C.), when the less exacting viewpoint held sway, to the era of the Stoics (a school of thought founded by Zeno of Citium a few decades after the death of Plato) when rigid insistence on identical cycles became, once again, de rigueur.

Plato, arguably one of the most original and influential thinkers of all time, was himself a great admirer of Pythagoras and his school. In contrast to Pythagoras, though, Plato was far more concerned with the mundane realm of human government and self-expression than with the abstract world of numerics. Consequently, it is no wonder that Plato refused to believe in *exact* reincarnation; precise repetition of events would negate the concept of free will, and without free will, the notion of representative democracy would carry little weight. Instead, he considered the human soul to be unique, sacred and transient, freely existing against the backdrop of an eternal, periodic cosmos.

Plato defined time as "the moving image of eternity," which he further identified with the "perfect circular motions" of the planets. Circles, he felt, represented the divine ideal, and the circular paths of the celestial spheres were the ultimate conduits of the godly essence. He went on to account for the various units of time, days, months, and years, as supreme expressions of the natural rhythms of creation: namely, the periodic motions of the sun and the moon. Finally, like the Chinese and Babylonians, he concluded that the Great Year, the time it takes for all of the planets to complete their revolutions simultaneously and return to their initial positions, was the ultimate cosmic cycle.

Plato's theory of history was based on his concept of the Great Year; he argued for endlessly repeating cycles lasting for thousands of years and consisting of successive intervals of order and chaos. The ordered phases, which he called "golden ages," represented periods of divine guidance and earthly harmony; in contrast, the chaotic phases were times when the gods withdrew from terrestrial affairs. These cycles of history wouldn't be identical, though, according to Plato; human lives wouldn't simply repeat themselves.

The Stoics, who carried Plato's views on cycles one step further, held that all cosmic and earthly events must precisely repeat themselves, again and again, for all eternity. In contrast to Plato's conception of human autonomy, they felt that free will was meaningless and history entirely predetermined. Each incident present in one cycle, no matter how insignificant, must be manifest in all; there would be no choice in the matter. For this reason, they urged complete acceptance of one's lot, no matter what happened—hence, the modern term *stoic*, meaning "dispassionate." A true Stoic would never flinch, even if he lost his whole fortune and his closest friends were killed; he would consider these circumstances necessary pieces of a puzzle preassembled by the gods.

When the Romans conquered Greek civilization, they incorporated much of Stoic philosophy into their system of belief. Thus, a Roman Stoic movement, led by prominent figures such as Marcus Aurelius (121–180 A.D.), the adopted son of Emperor Antoninus Pius, continued to uphold the banner of absolute acceptance of one's fate and complete disdain for ephemeral worldly pleasures, in light of the ever-fluctuating nature of reality. And, though Marcus succeeded his adopted father in the exalted position of Emperor of Rome, he never abandoned his principles and continued to live a life of utmost humility, eschewing many of the material trappings of his office.

Marcus's treatise, *Meditations*, written in the last years of his life, is an invective against vanity and pride in consid-

Endless Dance of Shiva

eration of the repetitive meaninglessness of the cosmos. Castigating himself and others for trying to hold on to the fleeting fancies of the mundane realm, he urges avoidance of all earthly attachments. His thoughts on cyclical time serve as a fitting summary of many of the prevailing themes of Stoic philosophy:

> All is the same: in experience, familiar; in time, ephemeral; in matter, sordid; and all in our days is as in the days of those we buried. All that now takes place took place in time past in exactly the same fashion; and doubt not the future will see the like. The plays are all the same; the cast only is changed!

So we see once again the hallmark of the "numerical" wing of Greco-Roman temporal philosophy: an insistence that the fate of the world is immutable and periodic, and that one must behave with great humility in deference to this fact. It is supremely ironic that the Stoics preached, on the one hand, that because of cosmic predetermination human behavior *cannot* be modified, while they advocated, on the other hand, that, in deference to cosmic predetermination, human behavior *must* be modified. What is the sense of trying to *convince* someone that his actions and opinions were already formed eons ago—as if he could change them if convinced? Nevertheless, in spite of these seeming contradictions, the Stoic view of cyclical time dominated Greek and Roman thought for centuries, until the rise of Christianity with its linear approach.

Leaving classical Greece and Rome, let's now further consider the Hindu notion of a cyclical cosmos, in many ways the most detailed conception of all. And in doing so we return full circle to India in our Dance of Shiva around the beliefs of the ancient world, a circuit that seems to reveal far more similarities than differences in these early civilizations' perceptions of time.

Wheels of Destiny

It is a tribute to the power of tradition that when India became independent from Britain in 1947 it chose an emblem that is over two millennia old to adorn its national flag. The so-called Wheel of Asoka, a dark, multispoked symbol of cyclical return, stands out sharply against the vivid red, white, and green stripes of the Indian banner as a telling reminder of an ancient faith's insistence on the transitory nature of all things.

Truly, it is paradoxical that Indian civilization, one of the oldest in the world, dating back thousands of years before the birth of Jesus, maintains such a strong identification with the notion of impermanence. One might think that such a long-standing culture, with its rich, complex history and well-traced family lineages, would embrace instead a worldview free of the possibility of radical change. Some historians have speculated that the highly changeable weather of India—torrential monsoons, tremendous winds, crop-destroying droughts—has played an essential role in the shaping of the Hindu mind-set, fostering the belief that all things must pass. (Note that, though we speak here of Hinduism, most of these notions are shared by Buddhists, Jainas and many other religious minorities in India as well.)

Whatever the cause, it is clear that the idea of cycles has remained such an integral part of life in India, even in modern times, that people there cannot plan out their private lives, conduct business, engage in leisure activities, or worship their deities without reckoning with its powerful force. The coiled serpent of recurrent destiny has wrapped itself around the Hindu spirit and has refused ever to let go, winding itself tighter and tighter as the centuries have progressed. And all the tools that modern technology has provided are ill equipped to ply it away.

Some have argued, therefore, that the Hindu preoccupation with cosmic repetition has resulted in an overall climate of stagnation; these critics contend that Hinduism, with its belief that the fruits of human effort are all illusory, provides little incentive for progress. This argument holds little water though; India's contributions to art, music, architecture, and science are substantial and certainly well respected.

Truly, Hinduism does seem to have provided the benefit of a certain measure of serenity for the people of India. There is much to be said for the comfort furnished by knowing that whatever disasters befall you or your country, they are due to be reversed someday when the circle of fate turns round again. Moreover, the Hindu creed provides devotees with a sense that the lives they have led could never be deemed a total failure, because all perceived mistakes and shortcomings would be smoothed out over time, after a number of cycles. Furthermore, unlike the Stoics, the Hindus don't believe in exact repetition of events. According to their scriptures, there is always the possibility for redemption in a future lifetime. Thus the Hindu veneration of cycles, while representing the source of stagnation for some, embodies the hope of divine restitution for others.

The cyclical tradition in India dates at least as far back as the writing of the Vedas: the oldest of the Hindu sacred texts, composed during a thousand-year period that ended at approximately 700 B.C. (As discussed, archeological finds containing *ourobouros* motifs and images of Shiva suggest an even earlier origin for these beliefs.) In Sanskrit, the word *vedas* signifies "knowledge about everything," reflecting the view that the sacred texts are the ultimate source of authority about the origin and purpose of the cosmos, as well as the meaning of human life. They speak of tribute to Brahma, the Absolute, who created the universe *outside of time*, not within it, and suggest suitable means of worshipping Him and His eternal creation.

The Vedas are divided into four sections, three of which contain prayers, poems of devotion, and instructions on holy rituals. It is the fourth part of the Vedas, called the Upanishads, that forms the philosophical basis of Hinduism.

The Upanishads are composed of 108 separate tales, each commenting on a different aspect of human and godly existence, and containing numerous expressions of admonition such as "what is the good of enjoyment of desires when after a man has fed on them there is seen repeatedly his return here to Earth?"

One of the primary themes of the Upanishads is the idea of cyclical return of souls, known in Sanskrit as the "Wheel of Samsara." Here, the world is pictured as a continuously spinning wheel, conveying the spirits of its inhabitants in an endless journey from birth, to life, to death, and then rebirth. Souls are seen as engaging in a progressive migration from body to body, with each corporeal experience leading to further development of piety and wisdom.

Moreover, in the depiction presented in the Upanishads, the material content of the universe itself is seen as transitory and cyclical. Periodically, as pictured, the cosmos is completely obliterated, with all extant life destroyed. Then, after a time, the Earth begins afresh, with new civilizations, new inhabitants, new gods, and new laws.

What of the Vedas themselves; if the world is continually created and destroyed how could the knowledge contained within these works themselves be seen as eternal? To resolve this seeming contradiction, Hindus postulate that the Vedic texts are rewritten at the beginning of each cycle.

Endless Dance of Shiva

Guided by divine wisdom, these writings assume the same rudimentary form each time, and thus preserve the semblance of continuity.

How long do these cycles last? As we've seen in our examinations of other ancient cultures, speculations on the lengths of these periods vary considerably, ranging from half a century for the Aztecs and Mayans, to thousands of years for the classical Greeks, and millions of years for the Babylonians and Chinese. Remarkably, the Indians are comfortable with even larger numbers than these. Thinking of time in terms of geological and astronomical phenomena, rather than human occurrences, the Hindus contend that the overall periodicity of the universe is at least 311 trillion years.

The detailed nature of the Hindu time scheme is described in the Puranas, a series of educational stories set to writing during the middle of the first millennium, but probably created far earlier than that. These epic tales serve to comment on and illuminate the Vedic teachings, by depicting the gods engaged in miraculous activities that convey allegorical meaning. In the Vishnu Purana, for instance, a delineation of the life of Brahma is used to illustrate the nature of universal death and rebirth. By carefully tabulating the ages of Brahma, one can readily determine the time scales of the Hindu cyclical scheme. The following is the Hindu chronometry, according to the Puranas:

1 divine day	1 (ordinary) year
1 divine year	360 divine days
1 Yuga	1000–4000 divine years
1 Mahayuga	4 Yugas, or 4,320,000 years
1 Kalpa	1000 Mahayugas
1 day of Brahma	2 Kalpas, or 8,640,000,000 years
1 year of Brahma	360 days of Brahma
1 life of Brahma	100 years of Brahma, or 311 trillion years

According to the Hindu calendar we are currently in the fourth Yuga of the 28th Mahayuga of the current Kalpa (cosmic cycle). For each Mahayuga, the first Yuga is supposed to be a golden age, the second, somewhat more decadent, the third, even bleaker, and the fourth and last Yuga, an age of darkness and despair. Thus, in the scheme of things, we belong to a distinctly corrupt period, soon to be followed by a golden age.

Supposedly, the original Hindu chronometry consisted of intervals that were of much shorter duration, and predicted that the golden age would begin sometime around the first or second century B.C. In this earlier scheme, each Yuga was considered to last several thousand *ordinary* years, rather than divine years. The divine years were later proposed when, after the foreseen golden age had already arrived, India remained in a state of complete turmoil, with the Hindu faith temporarily on the decline. Rather than to sacrifice completely the prophesies of their calendar, the Hindus decided to amend it by introducing the divine year to be 360 ordinary years, and then resetting all time scales to the larger value. Thus, in the "corrected" scheme, the estimated age of the current cosmic cycle became four billion years, rather than the mere 12 million years of the old system, and the estimated lifetime of Brahma became 311 trillion years, instead of 864 billion years.

In contrast to the cyclical theories of other ancient civilizations, the Hindu model bears little relation to direct astronomical observation of the celestial bodies. The lengths of the Yugas, Mahayugas, and Kalpas have been determined, not by the periods of planetary orbits, but rather by the arbitrary ages assigned to the gods. Nevertheless, it is remarkable that—seemingly by pure coincidence—the

Endless Dance of Shiva 39

Hindu predictions come closest in magnitude to modern estimates of the age of the physical universe: billions of years, rather than millions or thousands.

Hindu mythology speaks in stark terms about the cataclysmic end of each cosmic cycle. According to the Puranas, immediately following each Kalpa, there is a period, called *Naimittik Pralaya*, of universal dissolution and chaos, supervised by none other than Shiva the Destroyer. Under the divine guidance of Shiva, the sun's rays leap out from the sky—at first causing a general drought, then drying up the world's water supply completely, and finally setting the whole Earth on fire. Then, after this general conflagration is finished at last, a new world emerges from the ashes—with completely different inhabitants and civilizations.

An even greater horror, if one could possibly be imagined, takes place, according to this chronology, at the end of the lifetime of Brahma. When Brahma dies, all of the matter and energy in the cosmos dissolves with him in a process known as *Prakrta Pralaya*, literally "dissolution into non-entity." Everything that has ever been created—Earth, moon, sun, stars and planets—decays into nothingness in an all-consuming universal catastrophe. Then finally, after a long "Brahman night," a new supreme god is born, along with an entirely new creation.

Thus, even the death of the supreme deity does not imply, in Hinduism, an absolute cessation of time. Rather, Brahman tradition teaches that the number of cycles of the cosmos is infinite, and that when a god is destroyed a new godhead emerges, again and again, for eternity. This everlasting succession of deities is beautifully depicted in the Puranas as an endless trail of ants filing by forever in a straight line, and the number of universes in time is measured as "even greater than the quantity of all the grains of sand in the world." And, in a metaphor surprisingly similar to a recently proposed cosmological theory in theoretical physics (to be discussed in Chapter Eleven), cosmic emanation is described as a process of bubble creation: "like delicate boats they [the universes] float on the fathomless, pure waters that form the body of Vishnu. Out of every hair-pore of that body a universe bubbles and breaks."

In light of this extraordinary model of a perpetual cosmos, one can readily understand the tremendous popularity in India of the Dance of Shiva ritual; it symbolizes perfectly the ceaseless progression of the Kalpas, the constant turning of the Wheel of Samsara, the endless march of the parade of souls, the continual begetting and decimation of the gods, the ourobouric self-devouring and re-emanation of the universe—in short, all of the fantastic elements present in Hindu cosmology.

Nothing New Under the Sun

An Eastern mystic thinks of time and imagines cycles; a Westerner thinks of time and pictures a unidirectional flow. Today in the West it is hard for us to envision the periodic return of the cosmos; we are more likely to contemplate either a universe born in a solitary moment of creation or one that has always been here. Much of this Western bias in favor of a linear approach to time stems from the unequivocal exhortations of Judaism and Christianity that human history has an origin, sequence, purpose and destination, following an irreversible course of development explicitly mapped out in the Bible. While the Old Testament proclaims, in an apparent argument for cyclical transformation, that "there is nothing new under the sun," it also emphatically teaches that there was a single creation of the universe, and before that moment there was nothing. And the New Testament asserts that there is a linear pattern of events leading from the birth of Christ, to his death and resurrection, and finally to his Second Coming—a sequence that would make little sense as a continuous cycle.

During the long somber years of the Middle Ages, this choice of time models, linear or cyclical, was simply not a matter of personal preference. The Church condemned as heretics all who believed in the notion that time will repeat itself; those refusing to abandon their Greek or Eastern temporal viewpoints were subject to all the punishments wrought by blasphemy. Furthermore, St. Augustine's famous declaration, "Christ has risen but once," was taken as a strong admonition against the idea of periodic return; advocating an infinitely repeated resurrection of the Saviour would have been considered supreme heresy.

In the centuries following the Renaissance and Enlightenment (i.e., from the 17th to 19th centuries) the situation for Western advocates of cyclical time hardly improved. Soon after the Church's domination of philosophical discourse in Europe began to lift, science stepped in with its own linear theories of history and the cosmos, based on notions of continual human and scientific progress. Great thinkers of the age, from Francis Bacon to Herbert Spencer, generally advocated political theories arguing that human history can only lead to greater and greater achievement; i.e., history is irreversible, not cyclical, in nature. And the geological and biological sciences seemed to follow suit in advancing unidirectional models of natural change. From the terrestrial transformation proposals (stating that mountains and other features on Earth undergo erosion) of James Hutton and Charles Lyell to the evolutionary theories of Charles Darwin, natural scientists argued that the Earth and its denizens are continually developing. The world, in short, came to be seen as resolutely mortal—aging, in fact—with a fiery birth sometime in the past, a youthful era of organic development culminating in the emergence of present-day humankind, and a dusty demise far off in the future. Obviously, this vision is, by any standard, a far cry from the Eastern and classical notions of periodic creation and destruction.

Only in the past hundred years has the situation dramatically improved for those who wish for an eternal return of cosmic events and hope that the physical universe is immortal. At the close of the 19th century, two innovative thinkers, the German philosopher Friedrich Nietzsche and the French mathematician Henri Poincaré, independently developed physical models of an ever-recurring cosmos (though Nietzsche's approach is more intuitive than scientific, and Poincaré's theory applies only if the universe is a closed, finite system).

In the 20th century, numerous theories of an oscillating (cyclical) universe have been proposed—based on the carefully tested principles of Albert Einstein's general relativity. While theorists have continued to construct increasingly realistic models of an endlessly repeating cosmos, experimental physicists and astronomers have been hard at work measuring the physical constants that would determine whether or not any of these models are valid. And in the past few years, the old Eastern views of time seem to have made a considerable comeback, at least in spirit. In *A Brief History of Time*, the world renowned Cambridge physicist Stephen Hawking details a cosmological proposal similar in tenor to the idea of *ourobouros*: namely a universe that devours and reproduces itself. As mentioned briefly, another very recent cyclical approach, the bubble model, bears an astonishing resemblance (metaphorical, of course, not numerical) to the picture painted in the Hindu Puranas of the "ever-bubbling waters" of eternal creation.

In an interview published during the 1970s in the *Intellectual Digest*, noted Princeton astronomer, Jeremiah Ostriker, and three other scientists commented on the increasing parallels between the traditional Eastern approach and some of the modern physical views of time. The interviewer asked these scientists to explain why "words such as 'birth,' 'death,' and 'rebirth,' and phrases such as 'cycles of creation and destruction'" are occurring more and more in modern physics. Ostriker replied:

> There are a lot of similarities between the mystic's view of the world and Einstein's. I don't know whether it's coincidental that currently the best cosmology is a 'big bang' cosmology and that the best rival is a cyclic one, which is more like the Eastern view. I am intrigued. I suspect I could learn a lot just by thinking and talking about it.[4]

If scientists were to prove that, as the ancients believed, the universe is indeed recurrent, the philosophical consequences would be enormous—particularly if it were a proof of *exact* repetition. Imagine what it would be like to live with the knowledge that everything you've ever said or done will be exactly repeated, over and over again, for all eternity. How would it feel to know that one's very thoughts are carbon copies of a timeless script?

In terms of the intriguing issue of the impact of eternal return on human resolve, no modern philosopher has contributed more than Friedrich Nietzsche. One might even say he lived this question. Though Nietzsche's knowledge of physics was spotty, his heartfelt instincts led him to a passionate understanding of the profound dilemmas posed by time cycles.

ETERNAL RETURN
THE MIRROR OF CHANCE

> *I come again with this sun, with this eagle, with this serpent—not to a new life or a better life or a similar life: I come again and again in all eternity to this identical and selfsame life, so that I may again teach the Eternal Return of all things.*
> -Friedrich Nietzsche, "Thus Spoke Zarathustra"

Nietzsche's Encounter with Eternity

There is something about a walk through the woods that inspires a feeling of connection with the eternal. The timeless sweet fragrance of a forest breeze, the vaulting splendor of a grove of pines, the undisturbed tranquility of an alpine meadow, the uniform blue waters of a sylvan lake, all convey a sense of serene, everlasting beauty that cannot fail to induce awe and reverence.

When Nietzsche wandered along the wooded paths near Lake Silvaplana, in August, 1881, he was at a turning point in his life. Almost 37 and never married (unusual for his day; his one marriage proposal was refused), lonely, confused, plagued by ill health, and bored with his career, his sylvan wanderings were more than matched in turbulent complexity by his meandering thoughts and restless feelings. He was pining for something—anything—to which he could latch on for strength and comfort; he sought that intangible sense of purpose that had thus far eluded him.

Upon rounding a bend, he was captivated by the sight of a huge rock, towering above him like a pyramid. Suddenly, an astonishing thought "struck him like a thunderbolt" (as he put it later in his autobiographical account): that everything that ever transpires on Earth must occur again and again for eternity. Perhaps it was the ageless quality of the majestic rock and other natural surroundings that drew him to this staggering conclusion of perpetual recurrence. Whatever the cause of the sudden burst of inspiration, from that point on he was to extol the idea of *eternal return* in all of his significant writings.

Nietzsche drew new life from his abrupt realization. Although at first he was horrified by the notion that he would have to repeat all his moments of intellectual frustration, days of desperate confusion, and years of solitary anguish over and over again forever, he soon came to accept what he saw as the inevitable, and learned to tolerate the idea of cycles. Gradually, he even began to relish the stamp of durability wrought by eternal return; he proudly started to imagine himself as a permanent fixture of creation, repeatedly returning to the world as a prophet of recurrence. By winter 1881 he was downright joyous, refreshed after a long vacation in Genoa, and heady with the implications of what he felt was his new, monumentally important contribution to world philosophy.

The results of this excitement were the beginnings of Nietzsche's most famous book, *Thus Spoke Zarathustra*, an allegorical work that includes eternal return as one of its principal ideas. In preparation for his magnum opus, Nietzsche drew upon his considerable knowledge of classical and Eastern scripts, including the writings of Heraclitus and Empedocles, the sacred texts of Hinduism, and especially the mythological documents of the Persians. The result is a relatively modern book that has a time-worn "feel" to it, peppered with numerous references to primitive symbols such as "the ring of rings" and "coiled serpents of eternity."

Considering Nietzsche's interest and background in the study of ancient beliefs and cultures, it is not surprising that he named his epic work after a historical religious figure. The original Zarathustra was the founder of Zoroastrianism, an ancient Iranian cyclical-time-based religion; Nietzsche "borrows" him to be the main character of his book. Nietzsche's Zarathustra, however, serves mainly as a conduit of the German philosopher's own atheistic philosophy and bears little resemblance to his sixth century B.C. Persian namesake.

The principal theme of *Thus Spoke Zarathustra* is the idea of an enlightened man (Zarathustra) coming to grips with his own destiny in the face of endless recurrence and the nonexistence of God. Realizing that mankind cannot rely on God anymore to shape its fate, the protagonist preaches the doctrine that a superior human being will soon rise up and claim God's place as leader. This so-called superman will be so much in command of his emotions that he will view eternal return in a completely dispassionate manner (much like Shiva's measured indifference during his dance). And, compared to ordinary humans, with their petty squabbles, enormous greed, and limitless self-pity, the superman will seem a paragon of virtue, honor, and courage.

Upon completion of *Zarathustra*, Nietzsche embarked on a research project to prove scientifically what he already fervently believed: the doctrine of cycles. Though Nietzsche clearly was outside of his realm in these technical studies, he fancied himself a Renaissance man, with the same range of talents as Leibniz and Goethe (whom he admired for their breadth of knowledge), and wanted to demonstrate his own broad base of understanding in the sciences. Therefore, at this time he read a considerable amount of physics, albeit in popularized form, and set out to utilize the laws of classical mechanics in an original formulation of his recurrence theory. Although in his own lifetime Nietzsche never saw publication of his results, embryonic versions of his proof have appeared in posthumous collections of his writings.

Basically, Nietzsche's proof of eternal return employs the well-known principle of the conservation of energy. This natural law states that, in a closed system, total energy can never be created nor destroyed. He couples this law with the hypothesis that the number of particles, and particle states, in the universe is finite to show that the set of all possible physical events in the world must eventually exhaust itself (and thereby repeat itself). Or, in Nietzsche's words:

> The extent of universal energy is limited. Consequently the number of states, changes, combinations and evolutions of this energy, although it may be enormous and practically incalculable, is at any rate definite and not unlimited (and therefore) the present process of evolution must be a repetition. Inasmuch as the entire state of all forces continually returns, everything has existed an infinite number of times.[1]

Nietzsche's argument is readily understood in terms of simple family economics, with cash flow representing energy exchange. Imagine that a mother of three children, named Alvin, Betty, and Carl sets aside exactly $20, in total, for her children's allowance each week; this is her "conservation of money principle." Each of her children is very demanding; so to allocate the funds she randomly distributes a fixed sum of money (in dollar bills, not coins) to each child, such that the total cash allotted each week adds up to $20. Thus, for instance, $7 might be doled out to Alvin, $7 to Betty and $6 to Carl. Or, an allocation of $18 might go to Alvin, $1 to Betty, and $1 to Carl. Any distribution of whole dollars will suit as long as the fixed total is maintained. (Note that this situation is analogous to the distribution of energy to particles; if there is a finite amount of energy and a finite number of particles, then there are limits on the amount of energy each particle can receive.)

Now, the number of ways that the mother can dole out these allowances to three children is clearly a finite quantity; in fact, there are only 171 possible combinations. Once this figure is exhausted—after approximately three and a half years—then she must necessarily repeat herself and give out the same amounts that she has given to her children before ($7 to Alvin again, $7 once more to Betty, and $6 to Carl, for example). Thus one can say that there has been an eternal return of allowances; the children periodically receive exactly the same allocation.

Replacing dollars with energy units and children with particles, this is roughly the same line of reasoning that Nietzsche used to illustrate eternal recurrence. Once each of the possible energy distributions is realized (and possible particle positions and velocities for each energy level exhausted) then there must necessarily be an exact repetition of configurations. In other words, according to Nietzsche, conservation of energy dictates that all of the particles in the universe must eventually return to any given arrangement.

Because he was neither a physicist nor a mathematician, Nietzsche's argument was not very precisely stated. It failed to address many of the pressing conceptual issues of the day (such as the law of increasing entropy), and, consequently, it made no discernable impact on the history of science. Few scientists today are even aware of the German philosopher's methodical attempt to change our view of time from linear to cyclical.

Needless to say, Nietzsche's dream to be the prophet of a new worldview was never realized; he went mad in 1889 (only six years after the publication of *Zarathustra*), spent the last years of his life in an asylum, and died in 1900. The fact of his insanity, coupled with an unfortunate posthumous expropriation and distortion of his work by the Nazis, sullied Nietzsche's reputation for decades. Only now are his brilliant but flawed endeavors respected on their own merits by historians.

The Atomism Debate

Interestingly, at the same time that Nietzsche was constructing his makeshift scientific arguments for eternal return, a number of well-respected mathematicians and physicists were independently pursuing the same question and formulating their own theories of recurrence—as expected, in a far more rigorous fashion. To understand their lines of reasoning, one must first examine one of the major puzzles of 19th-century physics: the apparent contradiction between the two systems of thought represented by Newtonian mechanics and Clausian thermodynamics.

Newtonian mechanics is the study of physical objects and their interactions (such as gravitational attraction). Basically, the axioms of Newton dictate that a body, once in motion, must continue forever in a straight line at the same speed, unless affected by another body. This is the famous

Figure 5. Sir Isaac Newton, 1642–1727. (Courtesy of AIP Niels Bohr Library, W. F. Meggers Collection.)

law of inertia, the principle that guarantees that when an astronaut throws a monkey wrench into space, it proceeds indefinitely along a straight path without need for further propulsion.

Now suppose an object *does* become influenced in some way by a force. Then, according to Newton, the law of inertia no longer holds, and the object must change how it is traveling, either by speeding up, slowing down, or turning. All three of these possibilities fall under the general category of acceleration; thus a *law of acceleration* can be formulated, demonstrating the simple proportionality between acceleration and force.

For mundane objects (much larger than an atom), at ordinary speeds, Newton's laws are said to be exact (for smaller particles, scientists employ modern quantum theory). Thus, in theory, by applying these equations of motion to any set of particles, their future paths and speeds can be precisely determined. The application of this precept, called *mechanical determinism*, taken to its logical extreme, means that, given absolute knowledge, of the locations and velocities of all objects in the world at any specific time, then, in principle, the entire future of the Earth from that point on could be completely predicted. In practice, of course, this application would be virtually impossible; nevertheless, it boggles the mind even to imagine such powers of prediction.

Another important mathematical property of Newton's equations is their manifest *time reversibility*; that is, in the absence of friction, the particle motions derived from these laws appear exactly the same backward or forward in time. For example, imagine you are videotaping a small girl while she is playing a game of marbles with red and blue glass spheres of the same size and weight. Suppose you start the camera the instant after she has aimed and released a red marble in the direction of a stationary blue one. Following Newton's laws for elastic (frictionless) collisions, the red marble must roll on the floor in a straight line toward the blue, hit it, and stop. This, in turn, must cause the blue marble to move in a straight line away from its original position.

Suppose you now remove the videotape of the game and place it in your VCR. Because of time reversibility, the resulting program, except for the colors, should look the same played either forward or backward. Presumably, the forward version of the footage would appear exactly like the events just witnessed—the red marble released, then colliding with the blue.

Shown backward, however, the blue marble would appear to have been aimed at the red; it would seem to roll toward the red marble, hit it, and stop. This would appear to cause the red marble to deflect in a direct path towards the girl. Thus, qualitatively, the backward version would seem like a proper elastic collision in its own right, with the red and blue marbles simply exchanging roles. So, except for the color scheme, there would be absolutely no way to distinguish the two versions, no way of telling cause from effect. Thus, this marble scenario, in the absence of friction, would be completely time reversible.

Equivalent to both the principle of exact determinism and the concept of time reversibility is the law of *conservation of energy*. One can readily demonstrate that, assuming precise applicability of Newton's equations backward and forward in time, there can be no net gain or loss of mechanical energy in a given physical system. Thus, because we assumed, in our example, that the collision was perfectly time reversible, the set of marbles couldn't have lost energy. Otherwise, it would be possible to tell the two videotape presentations apart by noting a net loss of energy in the forward-time version and a net gain of energy in the backward-time version. In other words, to preclude distinctions between forward-time and backward-time versions, energy cannot be lost in either temporal direction.

In order for conservation of mechanical energy to hold strictly, there cannot be any dissipative (energy-reducing) forces such as friction. However, in any realistic interaction between average-sized material objects there must always be some friction involved; for example, when glass spheres collide, they must scrape each other for a brief instant, giving off heat energy in the process. Thus, a careful examination of the marble videotape, might, in fact, yield some evidence of energy loss (and time irreversibility) after all.

By contrast, if we consider particle interactions on the atomic or molecular levels (the collision of two hydrogen molecules, for instance), conservation of mechanical energy can be shown, in the Newtonian picture, to be valid in all cases. This is because the concept of friction doesn't exist on a tiny scale; friction is a product of the electrical interactions between *millions* of neighboring atoms, not just one or two elementary particles. Without the energy-draining effects of friction, collisions on an atomic level must thus perfectly obey the law of energy conservation.

Given what we have said about the deep connection between energy conservation and time reversibility, it follows, then, that all atomic and molecular processes must look exactly the same forward and backward in time. Thus, the complete reversibility found only approximately in the case of mundane, ordinary-sized objects, such as marbles or tennis balls, is *precisely* true for atoms and molecules. Physicists usually refer to objects of ordinary size as macroscopic bodies and objects of atomic size as microscopic bodies; hence, Newton's laws can be said to exhibit absolute microscopic, but not macroscopic, time reversibility.

The scientific belief in atoms is a relatively new one, little more than a century old (though the Greek philosopher Democritus had speculated on their existence). A gradual recognition of the profound differences between the microscopic and macroscopic worlds took place during the mid to late 19th century, when the modern theory of atoms was founded and the laws of thermodynamics (governing heat and energy transfer) were discovered. One of the leading figures in both of these endeavors, was the Austrian physicist Ludwig Boltzmann, the last of the great pre-quantum theorists.

Boltzmann was born in Vienna in 1844, the same year as Nietzsche, and lived for years in both Austria and Germany, alternating between the two countries. In different periods of his life, and at a time when it was much more common to change fields of discipline repeatedly, he worked in both theoretical and experimental physics. Moreover, he published articles in both physics and philosophy, and was considered extremely proficient in teaching as well as research. In short, he was a man of extraordinarily diverse talent.

Like many men of accomplishment, Boltzmann was willing to take substantial risks in his career—to stand up in support of unpopular notions. Unlike most of his contemporaries, he was a convinced atomist; that is, he believed that every natural element has a smallest constituent particle. Quantum physics in the 20th century was to prove him correct; but, in the times in which he worked, his scientific views were subject to a host of criticism from distinguished adversaries such as Ernst Mach and Wilhelm Ostwald, who each passionately argued against the existence of atoms.

Mach, who was very influential at the time both in science and society (as a well-respected physicist and, as well, a friend of many high-ranked Austrian politicians) vehemently asserted, in opposition to atomism, that all matter could be subdivided an infinite number of times and that there was no conclusive evidence for fundamental building blocks. Even when presented with substantial indirect proof of the existence of atoms, Mach refused even to consider it, claiming that he would only regard direct observational data.

In another attack on atomism, Ostwald, one of the founders of physical chemistry, argued, in a theory called "energetics," that energy was the only fundamental form of physical reality. Matter, he contended, was simply a particular arrangement of energetic substances and had no basic unit. In his zeal, he tried to rewrite basic chemistry books to include his concepts, spoke of extending his ideas to psychology by determining the energetics of true happiness, and even went so far as naming his house "Energie."

Boltzmann countered these eminent scientists' theories by pointing to the substantial physical and chemical evidence for atoms, stemming all the way back to John Dalton's formulation in 1803 of a table of atomic weights, and including other 19th-century achievements such as William Thomson's estimation of the number of individual particles in a cubic millimeter of water, James Clerk Maxwell's kinetic theory of matter, and Dmitry Mendeleev's periodic chart of the elements.

Boltzmann, however, in spite of numerous published articles and lectures, met with little success in convincing other physicists to recognize the merits of atomic theory; in fact, during the last few years of his life, he had to confront a growing interest in the ideas of Mach and Ostwald. When Boltzmann committed suicide in 1906, many of his close friends attributed it to his feelings of isolation, due, at least in part, to what he saw as a losing battle for atomism.

Ironically, only a few years after his suicide, Boltzmann's ideas were vindicated. A series of important experiments paved the way for a new form of atomism that became almost universally embraced: namely, the quantum theory of matter and energy. While Mach and Ostwald's models were relegated to the "dustbin of history," Boltzmann's approach (with the addition of quantum mechanical terms) proved to be the best overall picture of the microscopic workings of natural substances. Today, few would doubt the fact that all materials are composed of myriad atoms, with each representing a particular element of the substance.

Time's Arrow

With atomic theory verified, the fact of time reversibility for most ordinary interactions on the microscopic scale has been completely confirmed (some physicists have claimed, though, to have found peculiar cases of microscopic time reversal violation for certain special forces). Clearly, Newton's laws dictate that atomic motions must appear exactly the same forward and backward in time.

When viewing the behavior of a material object on a macroscopic scale, however, a strikingly different picture emerges: manifest time *irreversibility*. An ice cube, for instance, taken as a whole, clearly must melt in an irreversible fashion, though the water molecules within it seem to engage in reversible scattering patterns. It is thus a truly perplexing result of 19th-century physics, that one can vary the scale of observation and obtain such disparate behavior. Yet the fact is clear: macroscopic systems possess unidirectional arrows of time determined by the law of increasing entropy.

The theoretical concept of entropy (from the Greek word *tropos*, meaning "transformation"), a measure of the disorder of a physical configuration, dates back to the work of Rudolf Clausius in the mid-19th century, who, in turn, derived many of his conclusions from the pioneering experimental investigations of Sadi Carnot in the early 19th century. While investigating the efficiency (work produced per energy utilized) of steam engines, Carnot found that, though the *total* quantity of energy in a closed process must remain constant, the amount of energy *usable* for mechanical work generally decreases during a given production cycle. (Steam engines operate by heating up water to create steam, using the steam to drive a turbine, and finally expelling the vapor into a cold water reservoir—thus relying on the temperature distinction between the hot and cold substances to generate work.) Carnot further determined that even the most efficient steam engines cannot convert all of their intake energy into usable work; at least a portion must be wasted as heat exhaust.

In 1865, Clausius restated Carnot's results in the form of a broader statement involving the concept of entropy, which he defined in terms of the heat exchanged during a given thermodynamic process and the temperatures of the objects involved in the heat transfer. The law of entropy, also called the Second Law of Thermodynamics (the First Law is the principle of energy conservation), can then be expressed in a number of equivalent ways.

One statement of the Second Law, for instance, is that heat can never spontaneously flow from a cold object to a hot object; it must always travel from hot to cold. Thus hot objects naturally become cooler and cool objects hotter; the reverse process would require an artificial infusion of heat. The end result is that all isolated processes tend toward thermal equilibrium (the same temperature throughout the whole system).

Another formulation of the Second Law makes its connection with cosmology clearer. It states that the total entropy of a closed system (presumably including the universe in general) invariably tends toward a maximum value; that is, it never decreases. Because entropy is a measure of the lack of availability of energy to do work, this means that, as time goes on, there is less and less usable power to be had.

And because this law applies to all physical systems, there are astronomical implications as well: solar engines eventually run down.

Solar energy, the primary source of power on Earth, via solar heat and sunlight, often seems inexhaustible; yet the Second Law of Thermodynamics and stellar theory provide unmistakable evidence that the sun, after a series of catastrophic transformations, will run out of fuel and turn into a burnt-out carcass billions of years from now.

This frightening chain of events will be initiated when the sun's primary mechanism for power generation—the fusion (combining) of hydrogen into helium in a massive, heat-releasing chain reaction—begins to slow down considerably after most available hydrogen fuel is expended. At that point, the solar core, the enormously hot center of the sun where most of these nuclear reactions take place, will lose most of its internal heat-generated pressure and nothing will stop it from gravitational collapse due to its own weight. Like a hot air balloon gone cold and shrinking, the core will become smaller and smaller, folding up into itself with little hindrance.

Paradoxically, though, once the core contracts sufficiently, the mere act of collapse will cause it to generate more energy, this time from shrinkage, rather than from hydrogen burning. As the core implodes, its temperature will rise again, temporarily, and it will expel large quantities of energy. Then, the pressure caused by the excess core energy will push the loosely bound, gaseous material surrounding the center outward in a tremendous burst. At that point, this fiery halo will glow red and expand until it occupies a region equivalent to the Earth's orbit, engulfing the Earth, Venus and Mercury in the process. After this shrouding of its planets occurs, the sun will no longer be a yellow, average-sized star, but, rather, an incredibly voluminous red giant.

The final stages in the sun's evolution will take place when the sun's gravitation is no longer sufficient to bind together the red-glowing gases in its halo. Consequently, these peripheral materials will eventually dilute and scatter, exposing the hot white core, all that is left of the sun, to deep space. This tiny, extremely dense (a teaspoon of its material weighs more than a car) remnant, known as a white dwarf, will glow for a time, burn off the last of its fuel, and finally become a cold dark relic, called a black dwarf, as the sun's history draws to a close.

A far more turbulent fate would await the sun if it were at least 50 percent more massive, in which case it would undergo a much more energetic collapse. During the last stage of its evolution a supernova explosion (such as the one that produced the Crab Nebula) would take place, blasting off most of its material and leaving only a rapidly shrinking core. The end result would be a far more compact remnant, called a neutron star, composed of pure neutronium (highly compressed neutron matter), so dense that a mere thimbleful would weigh more than all of the cars in the world put together. The neutron star would be even less energetic, however, than a white dwarf, emitting so little radiation that it would be practically undetectable.

And if the sun were even heavier to start out (two or three times its actual mass today), its final state would be even more bizarre: a remnant even denser than a neutron star. In this event, the core would collapse with such intensity that its constituent material would be absolutely pulverized; all elementary particles within (neutrons, protons, etc.) would be crushed beyond recognition. Finally, a black hole would be formed, an object so compressed and, hence, gravitationally powerful that not even light can escape.

Because light can't emanate from their surfaces, black holes are, by nature, completely invisible—the coldest and darkest objects known to science. Lethal to any objects that might encounter them, they gobble up all matter or energy in their vicinity and grow larger in the process of accumulation, occasionally even cannibalizing each other. They truly represent the definitive cosmic "vacuum cleaners," eradicating all things unfortunate enough to be in their way.

Black dwarfs, neutron stars, and black holes are the ultimate products of star death. Their existence in the cosmos confirms the Second Law of Thermodynamics as applied to the stars. The law of increasing entropy mandates that usable energy must either stay the same or decrease as time goes on, and, in fact, the energetic output of a dim stellar end-state, such as a black dwarf, is far, far less than that of a vibrant, relatively young star, such as the sun.

It is clear, then, why scientists often refer to entropy as "time's arrow"; any natural progression of large-scale events proceeds in the direction of greater, rather than lesser, entropy. The aging of stars, for instance, provides an irrefutable portrait of time's ever-flowing stream; a film taken of yellow stars evolving into black dwarfs would only make sense viewed forward, not backward, in time. This macroscopic irreversibility is in marked contrast to the microscopic reversibility of the atomic interactions taking place within the stars; thus, this presents another manifest example of the dependence of objects' apparent temporal behavior on the scale in which they are viewed. Yet, in spite of the seemingly time-invariant motions of the particles within them, the long-term behavior of stars as a whole furnishes an unmistakable portrayal of nature's steady, unidirectional erosion.

Heat Death

Given our knowledge of the ultimate fates of stars, we are left with a rather bleak portrait of the future of the cosmos—assuming, that is, that the law of entropy increase is strictly true for the whole universe, not just, say, for our particular region. (As we'll see, the assumption that the Second Law is universal has been a source of considerable controversy, but let's, for the moment, suppose this to be the case.) The universe, then, it would seem, will be doomed to reach a state of maximum entropy: the so-called heat death.

The concept of heat death was first made explicit in the 1850s by the German physicist Hermann von Helmholtz, as an extension of the theories of Clausius. Helmholtz, who strongly influenced the scientific thinking of Boltzmann, applied the Second Law to predict that eventually the universe will reach a state of uniform temperature, upon which it would be impossible to generate any usable energy and all natural activities would necessarily cease; as Helmholtz related in an 1854 lecture on the subject: "the universe from that time forward would be condemned to a state of eternal rest."

Astrophysically speaking, heat death would mean the complete conversion of viable stellar matter and energy into inanimate remnant material. First, all stars would cool down and, depending on their mass, turn into either black dwarfs, neutron stars, or black holes. At this point, many of the black holes would begin to gobble up what free matter remains—planets, asteroids, interstellar gases—as well as cannibalize other black holes. Finally, all that is left of the universe would be a collection of black holes and perhaps a small quantity of other forms of debris. Cold and lifeless, this shadowy array would represent the final phase of the cosmos and, according to some theories, the end of time.

Because of the unsettling nature of these prospects, many physicists have questioned the assumptions on which they are based. One of the most cogent challenges to the idea of heat death was presented by the distinguished British physicist E. A. Milne (not to be confused with the author of *Winnie the Pooh*, A. A. Milne, who had nothing to do with this issue!) at a 1931 meeting to discuss the evolution of the universe. Milne argued that the Second Law of Thermodynamics could not be applied to the cosmos as a whole, but rather only to isolated segments. Therefore, he pointed out, universal heat death was far from inevitable.

Milne's proof rests on the fact that the Second Law holds only for processes in which division into two parts, an inside and outside, is possible. For example, in the case of a steam engine, work can be produced, and thus entropy increase occur, only when heat is expelled to some external reservoir such as a tank of cold water; otherwise, the Second Law doesn't directly apply. Now, for the universe as a whole, no outside could be defined, only a vast inside; thus the Second Law needn't be universally valid and heat death wouldn't necessarily take place, Milne argued.

Milne was not the first critic of heat death. Many decades earlier Boltzmann, in spite of his great respect for Helmholtz, argued vehemently against its inevitability, using atomic theory to make his case. Boltzmann was working in the context of attempting to explain irreversible, macroscopic thermodynamics in terms of reversible, microscopic atomic motions. His achievements in this area undoubtedly brought him his greatest acclaim, albeit a mostly posthumous recognition of the profound importance of his work.

The Recurrence Controversy

If one visits the Central Cemetery in Vienna, one might be struck by the stunning white marble bust of Boltzmann that sits on top of his tomb. Examining the gravesite closely, one notices that carved into the tombstone is a mathematical formula relating the entropy of a substance to its internal atomic configuration, an expression that is arguably Boltzmann's most consequential scientific achievement. Boltzmann's equation expresses the entropy of a given macroscopic body at a particular temperature (a glass of hot water, for instance) as a direct function of the number of ways it can be constructed out of microscopic particles (water molecules, for example). His results represent the culmination of years of effort in trying to resolve the seeming conflict between large- and small-scale behaviors (time-irreversibility versus time reversibility) of physical systems.

As a strong believer in atomism, Boltzmann was highly troubled by the fact that macroscopic thermodynamic mechanisms, such as heat passing from hot air to cold air, could not, at that time, be codified in terms of the microscopic Newtonian dynamics of the air molecules themselves. To him, any self-consistent atomic theory must define quantities extending throughout a substance, such as overall temperature and entropy, in terms of the specific positions and velocities of the particles constituting that material. Otherwise there would be two theories—one for the large, the other for the small—and this would not do.

After working for a while on this issue, Boltzmann found what he thought was an adequate solution to the problem. He defined a statistical quantity, called H, based solely on the microscopic variables of a system, and showed that H possessed many of the same properties as the system's entropy. In particular, in his celebrated H-theorem, Boltzmann proved that, for an atomic system starting out with a large value of H, H must continually decrease until it reaches a minimum value. In other words, in a kind of reverse Second Law, H diminishes with time. Therefore, if one defines entropy in terms of the *absence* of H, then this theorem demonstrates why entropy tends to increase to a maximum amount. As H goes down, entropy goes up.

Boltzmann's theoretical results proved to be a major victory for atomism, and a clear first step along the route to reconciling large-scale and small-scale behaviors. Somehow out of the haphazard, ever-fluctuating motions of molecules, he managed to derive a quantity that was continuous and unidirectional in time. It seemed that, via Boltzmann, a comprehensive Newtonian proof of the Second Law of Thermodynamics would be soon at hand. Boltzmann's H function appeared to provide the missing link between thermodynamics and Newtonian mechanics.

But from the start there were considerable problems with Boltzmann's conclusions. For one thing, his proof rested on the *ad hoc* supposition that entropy began as a very small quantity (or, conversely, that H started out as very large). If entropy were, in fact, very large (close to its maximum value) at the time that the world began, all processes, according to Boltzmann, would tend toward the direction of smaller entropy as the Earth evolved. Obviously they haven't; so how can we explain the unmistakable fact that time's arrow points toward entropy increase, not decrease?

Given, then, the validity of the H-theorem, one is left with two options to overcome this serious shortcoming (that the H-theorem says nothing about entropy increase). The first possibility is to assume, as a given, that the early cosmos possessed little entropy. Because small values of entropy tend to grow larger, there was a natural increase from this minimum value to its present amount. Extrapolating toward the future, this scenario predicts that the universe must eventually reach heat death, when the overall entropy finally obtains its maximum quantity.

But Boltzmann chose instead a second option, one that demonstrated his disdain for the bleak idea of cosmic demise. Rather than making any suppositions at all about the uniformity, at any given time, of the entropy content of the universe, he presumed that the cosmos is a mixture of both low-entropy and high-entropy pockets, with the former evolving toward greater entropy, and the latter toward lesser entropy. The natural history of our own region of the universe and the fact that time's arrow points forward, then, are simply consequences of the circumstantial detail that we happen to reside in a zone of increasing entropy. So according to these speculations, if we were in another, entropy-decreasing part of the galaxy, clocks would run backward, chickens would be "unhatched" into eggs, and toast would reemerge from a toaster as soft bread.

Boltzmann realized, of course, that such projections seemed highly fanciful; there wasn't a single piece of evidence to demonstrate that the universe has regional arrows of time. Yet because he was wed to the premise that time's flow stems from entropy increase, he was stuck with his rather questionable hypothesis of a patchwork cosmos.

These were not the only troubles that Boltzmann had to endure to defend his theory. Another serious criticism of his work, the so-called *recurrence paradox*, was even more of a match for his keen argumentative skills. Raised by the

Eternal Return

mathematician Ernst Zermelo in 1896, this formidable objection to the H-theorem was to keep Boltzmann busy during his last decade of life.

The recurrence paradox relies on a startling conclusion by Henri Poincaré that turns out to be remarkably similar to the amateur speculations of Nietzsche. Basically, Poincaré's results state that any closed system, containing a finite number of atoms, must, after a time, return infinitesimally close to any given initial configuration. In another words, in a manner similar to Nietzsche's eternal return or the Hindu Kalpas, all finite atomic systems must repeat themselves again and again.

Zermelo used this idea of Poincaré cycles to argue vehemently against atomism. Surely, he argued, atomic systems couldn't maintain, albeit statistically, a unidirectional arrow of entropy increase (as in the H-theorem) while, at the same time, cycling periodically through a finite set of possible configurations. If this were true, then any state of maximum entropy, on either a local or universal scale, would necessarily be broken when the atoms constituting that state returned to a previous arrangement of lower entropy. Assuming this was the case, we would witness quite extraordinary behavior. For example, though the particles forming hot water and cold water might, at first, mix together in a process leading to higher entropy, eventually these molecules would cycle back to their original positions and velocities, separating the water once again into two different temperature regions, and thus decreasing the overall entropy back to its initial value. Clearly, this never would happen. Referring to this apparent paradox, Zermelo thus concluded that, because atomic theory seemed to violate the law of entropy increase, atomism must be invalid.

Figure 6. Henri Poincaré, 1854–1912. (From Images of Einstein Catalogue, New York: Center for History of Physics, American Institute of Physics, 1979. Courtesy of AIP Niels Bohr Library.)

Boltzmann's response to this challenge was to concede the existence of Poincaré cycles, but to point out that their lengths would be astronomical; it would take trillions and trillions of years for even the molecules forming a tiny grain of sand, let alone those constituting the whole universe, to exhibit a single Poincaré cycle of recurrence. So, Boltzmann asserted, while the irreversibility of the H-theorem (and the Second Law derived from atomic principles) was not strictly true for all eternity, for all intents and purposes it must be considered valid for any reasonable span of time.

Eternal Return Revisited

The ancients maintained that the cosmos must repeat itself. Could they be right? As Nietzsche, Poincaré, and Zermelo, among others, have pointed out, classical Newtonian physics, combined with the notion of atoms, dictates that any closed system must recur eternally. While this recurrence time can be shown, in virtually all examples, and especially in the case of the cosmos as a whole, to be astronomically long beyond all comprehension, still, over the course of eternity, in any bounded entity there surely is a finite chance of repetition. If you place a billion monkeys in front of a billion typewriters and wait a billion years, it is quite probable that one of them will type out a line from Shakespeare. Similarly, if you wait a long enough time—so it's argued—the myriad possible atomic patterns have a distinct likelihood of repeating themselves.

As the Hindu scriptures point out, the lifetime of Brahma, in terms of human years, is even greater than the number of grains of sand on Earth. We should bear such a quantity in mind while trying to picture the far lengthier durations required for Poincaré recurrence. Yet as lengthy as these periods might seem, we mustn't prematurely dismiss the idea of eternal return. Even if recurrence were to take a trillion years raised to the trillionth power, it would still embody an exact, or near-exact, repetition of cosmic events—a circumstance that would be fascinating, nevertheless, in spite of its enormous time of completion.

Perhaps the universe, after all, is just a game, one with hosts of "playing pieces" (the elementary particles) and innumerable possible "moves," (particle interactions). Is it ridiculous, then, to imagine this game repeating itself over and over again in the course of eternity? Consider the game of tic-tac-toe, with its small number of allowable moves; exact repetition of matches is assured after only a few hundred rounds. Chess would take much longer for all the possibilities to be exhausted. Still, chess games must eventually repeat themselves. Even the human genome is limited in its number of combinations. Thus, given enough time, an exact copy of the genetic code of any living organism would be reproduced by a mere roll of the biological dice. Might these same arguments apply to the physical universe as well?

Perhaps. But one factor that these 19th-century thinkers failed to take into account was the behavior of space itself; in the absence of evidence to the contrary, they naturally assumed that the dimensions of the universe were rigid. However, just a little over a decade after the death of Boltzmann, two astonishing discoveries were made in astronomy that shook the very foundations of our beliefs.

First, in 1924, the astronomer Edwin Hubble demonstrated that ours is not the only galaxy in the cosmos. Prior to that time, the terms "galaxy," "Milky Way," and "universe" were all considered synonymous; it was thought that the limits of one were the limits of the others. However, Hubble employed a powerful distance-measuring technique using regularly pulsating stars, called "Cepheid variables," to show conclusively that many of the hazy celestial objects once believed to be in the Milky Way, are, in fact, galaxies in their own right. (By knowing the pulsation rate and brightness of a Cepheid variable star, a simple formula can be used to determine its distance, and thus ascertain whether or not the star lies within the Milky Way.)

Once he established the huge distances of these other galaxies, Hubble made a second monumental discovery about the behavior of their velocities. By utilizing the so-called Doppler effect to measure the shifts in galactic spectral lines, he found evidence that, aside from our closest neighbors, all other galaxies are receding (moving away from the Milky Way). In the Doppler effect, the movement of objects, relative to an observer, causes their light spectra to alter in predictable ways. The spectra of objects moving away from an observer shift toward lower frequency colors such as red, and those of objects moving toward an observer shift toward higher frequency colors such as blue. In his telescopic measurements, Hubble found redshifts—meaning movement away from us—for all observable galaxies outside of our Local Group (the Local Group includes nearby galaxies, such as Andromeda). These shifts, along with other evidence, seemed to indicate that the universe is expanding, and that, at least for the near future, space itself is becoming more and more extensive.

The expansion of space has a direct bearing on the possibility of eternal recurrence. Let us return to our game analogy. If a playing field (a chessboard, for instance) continues to grow, and the pieces continue to have more room (squares) in which to operate, then, as there are more and more possible moves, it becomes less and less likely that the matches will be identically repeated. The same can be said for the cosmos; Nietzsche and Poincaré's arguments for recurrence do not seem to hold for ever-expanding systems.

Interestingly, the malleability of space does seem to allow for other forms of cosmic repetition. Although we now know that the universe is currently expanding, many scientists also believe there is a good chance that it will some day collapse. This universal contraction would lend itself to diverse possibilities for a recurrent cosmos, summoning up images reminiscent of the Babylonians, Indians, and Greeks of cyclic creation, destruction, and re-creation of the world.

PART 2

THE MODERN VIEW OF THE COSMOS

THE EXPANDING HEAVENS

The Seed grew into a body with the brilliance of a thousand suns and therein Brahma, who is the forerunner of everything, took form.
-*Manu Samhita (Hindu sacred text)*

Unholy Designs

Cosmology as philosophical speculation is thousands of years old. Throughout history, each of the Earth's peoples has devised a scheme by which it has attempted to discern the origins and destiny of the world, a means of trying to hoist a candle into the murky recesses of our mysterious beginnings and to shine a revelatory torch on the nebulous visage of our hidden fate. As if to illuminate an infinite cavern, each culture has tried to cast off the vast darkness of the unknown with its own beacon of boundless faith. Yet for all these bold attempts, over scores of centuries, the shadowy haze that surrounds the past and future has barely been penetrated by philosophical or religious inquiry.

Can science do any better? Clearly, it can. *Physical* cosmology is less than a hundred years old—far, far younger than its metaphysical cousin—and already it has unveiled a remarkably clear panorama of the natural history of the universe. Obviously astronomy can't show us everything, but we can be reasonably sure that the portraits it *has* revealed represent the results of painstakingly checked and double-checked measurements, each generated by state-of-the-art equipment and techniques.

Producing these detailed pictures of the heavens are telescopes of all shapes, sizes, and frequency ranges, descendants of the Pisan astronomer Galileo Galilei's unholy instrument that was the bane of the Catholic Church for many years. Galileo, who lived in Italy right before the time of Newton, committed the "blasphemous" act of trying to replace, as means for envisioning the world, the stained-glass-filtered illuminations of religious faith with the focused product of mirrors and lenses. For supposed transgressions related to astronomical discoveries that shook those with the orthodox view of the cosmos, Galileo was singled out and thoroughly condemned by the Church. It wasn't until over a century after his death that the Church finally admitted that Galileo's results were valid.

Galileo was not the inventor of the telescope. That distinction belongs to Dutch spectacle-maker Hans Lippershey, who, in 1608, stumbled upon its design by chance. As the story goes, Lippershey happened to place one lens in front of another, and looked through them toward a distant church steeple. He discovered, to his amazement, that the combined power of the lenses caused the image of a weathervane on top of the steeple to appear quite large before his eyes. Later he found that he could enclose these lenses linearly in a tube and produce a wide range of effects by altering the lens curvature and tube length. Thus, the first telescope was created.

Expanding Heavens

Word of this discovery reached Galileo when the "optical tube," as it was then called, was exhibited at an exposition in Venice. He soon set out to make his own device. First he ground a concave lens and placed it on one end of a lead pipe. Then he mounted a convex lens on the other end of the pipe (the viewing end). He found that by peering through the tube, distant images could be seen at nine times their unmagnified dimensions.

Over a period of several months, Galileo worked on improving the design of his telescope, with the intention of using it to make astronomical observations. He came to realize that by making small alterations in the structure and composition of the scope, he could increase its magnification by another factor of four. When he tested his device by aiming it at various celestial objects, a revolution in scientific understanding of the universe was begun.

The first body on which Galileo trained his optical instrument was the moon. He found to his sheer amazement that it was pocketed with jagged mountains, vast craters, spacious plateaus, and deep ravines; its topography was reminiscent of many regions of Earth. He further noticed that the dark side of the moon was dimly lit, a phenomenon that he correctly postulated was due to the reflection of the sun's light off the Earth's surface and onto the moon. From all this he concluded that the moon and the Earth share many common properties.

Following his discovery that the moon was a world in its own right, Galileo went on to examine other planets and their properties. Without exception, he found that they had visible features as well, appearing like small disks or crescents against a background of pointlike stars. He surmised that because the moving planets had noticeable breadth and the fixed stars didn't, the former had to be much closer to Earth than the latter.

When he turned his gaze upon Jupiter in particular, he saw that it was surrounded by four objects, which he called "Medicean stars," in honor of the influential Medici family of Florence. Careful observation, night after night, indicated that these bodies seemed to orbit Jupiter. He concluded that, like the Earth, Jupiter possessed moons.

In the case of Venus, Galileo found strong evidence that the planets revolve around the sun and reflect solar light as they do so. When he first fixed his telescope upon Venus, he observed its appearance to be that of a narrow crescent. But then, as time progressed, he noticed that this sliver expanded to become a wider crescent and then a disk. These phases seemed to indicate that the glow of Venus was caused, not by the planet's own light, but by light from the sun bouncing off its surface and then traveling to Earth. Furthermore, the diameter of Venus appeared to change significantly during these phase transitions. During the crescent phase, Venus seemed much wider than during the disk phase. This variation could be explained only by assuming that Venus orbits the sun.

It is impossible to overestimate the impact that these results had on the 17th-century European view of Earth's place in the cosmos. Prior to Galileo, most people in Europe subscribed to the Aristotelian view that the stars, planets, sun, and moon were luminous objects attached to fixed invisible spheres orbiting the Earth. This theory of celestial behavior stemmed, in part, from centuries of Greek (and Babylonian, Egyptian, etc.) thought about time cycles and the heavens, dating even further back than Pythagorean notions of the "music of the spheres."

Conservative and suspicious in its preachings, the Church had incorporated Aristotle's ideas, with some theological modifications (because, of course, Aristotle predated Christianity), into its own dogma. The bulk of European intellectuals, ignorant of alternate views of the cosmos or afraid of being labeled heretics, continued to adhere to the views of the Church. There were, however, some prominent exceptions; the Polish astronomer Nicholas Copernicus created quite a stir in 1543 when he suggested, in his book *On the Revolutions of Celestial Bodies*, the idea of a heliocentric (sun-centered) planetary system (Galileo, almost 90 years later, adopted this Copernican notion). But for the most part, few dared to deviate from the Church-sponsored Aristotelian dogma.

One might wonder how the Church could harbor essentially the ancient Greek view of natural cycles, with regard to the solar system, but reject this belief in the case of the universe as a whole. Unlike the Greeks, Babylonians, Hindus and other cyclical-time-based cultures, who believe that the repetitive motions of the planets portend a similar repetition of cosmic events. Catholics (as well as other Christians, Moslems, and Jews) have traditionally eschewed such arguments. According to orthodox Church doctrine, the cosmos was created by God at one fixed point in time as a fully functional system. In other words, the universe is like a ticking clock that was spontaneously brought into being with its stem already wound.

It is interesting to contrast the reaction that classical Greek scholars and Church theologians might have had upon first observing an enormous clock, ticking away in the sky with an exact rhythm, in analogy with the clockwork behavior of the planets. On the one hand, the Athenians might have concluded that this instrument was so perfect that it couldn't possibly run down; therefore, it must eventually reproduce its own pattern, and, by extension, time must repeat itself. On the other hand, the pious Christian clergymen might have resolved, on seeing this beautiful heavenly clock, that a divine being must have created such a fine mechanism and that such a being must have the power to someday take it away. Thus, the same regular behavior might be interpreted in either a cyclical or linear (creationist) fashion.

These purely metaphysical arguments were too obtuse for Galileo. He wanted to see for himself the composition and behavior of the universe, instead of drawing inferences based solely on logic or scripture. In this way, his style of thinking was a radical departure from the orthodox academic approach of his times, which warned against too much scientific experimentation and analysis of God's designs.

The Church tolerated Galileo at first, viewing his results as interesting curiosities to be taken as amusing alternatives to what was seen as biblical truth. But when Galileo began to insist that his findings about the nature of the planets (that they orbited the sun, for instance) *superseded* Church doctrine, that was simply too much for the authorities to endure.

Finally, strong action was taken by the Vatican against the Italian astronomer. He was tried and placed under house arrest for nine years, where he grew blind and ill. His writings were banned and ultimately he was forced to recant his ideas publicly, especially his belief in a heliocentric (sun-centered) cosmos. He died a broken man in 1642.

Galileo's revolution—of exploring the heavens by use of telescope—triggered a prolonged fight in Europe between the forces of scientific reason and religious dogma over who was to lay claim to absolute knowledge of the cosmos. In this prolonged bout between science and religion, round one was won by the Church. For decades after Galileo's death it was taboo for religious Catholics to discuss his ideas.

Round two, however, was won by science in a knockout punch. As newer and better telescopes were developed, Galileo's discovery that the Earth was but one of many extended solid bodies orbiting the sun became the overwhelmingly favored model, and was even accepted by the Church. (Over the years, the Church adopted a somewhat less literal interpretation of the Bible.) Moreover, because of organized religion's previous intransigence on this issue, it came to be seen as having little to do with cosmology. Gradually, in the West, experimental science and religion became decoupled, with the former assuming the dominant role in discourse about the cosmos and the latter relegated to the more reflective domain of ritual practice and meditative experience.

But there may indeed be a third round of this "match." An increasingly large segment of the scientific community is now beginning to discuss God openly and unapologetically in its literature. Perhaps this is due to an increasing appreciation in the West for Eastern religion and philosophy—which never became dominated by pure science—or maybe it is caused by a growing realization that we are pressing up against the very limits of the scientifically knowable. Books such as *God and the New Physics*, by Paul Davies, *Reading the Mind of God*, by James Trefil, and *The Tao of Physics*, by Fritjof Capra, have purported to find deep connections between physics and religion. Some Western scientists now speak freely about determining God's role in choosing our universe from the set of all possible cosmologies. Quite conceivably, these are harbingers of a new era in which the religious community once again plays a significant role in scholarly cosmological discourse.

Skygazers

In the centuries that followed Galileo's remarkable discoveries, a host of innovations improved the ability of astronomers to survey the cosmos by telescope. And, with each improvement, more and more became known about the vast reaches of space that surround us.

Newton, who was born on the date of Galileo's death, shared his scientific predecessor's fascination with telescopes. After experimenting with various telescope designs, Newton became particularly interested in increasing the precision of these astronomical devices. His considerable experience with prisms taught him that lenses tend to reflect colors at various different angles. He saw this as a major drawback of lens-based telescopes such as Galileo's. Because of the different paths traveled by light of different hues, it is impossible for the entire spectrum of colors to be brought together by a lens to a single focus. Newton suggested that this phenomenon, called *chromatic aberration*, could be avoided by replacing the lens at the end of the tube with a highly reflective curved mirror. Then, by peering into the eyepiece (front lens), which magnifies the tiny image, one could observe a relatively undistorted view of the light focused by the mirror. Newton's mirror telescope derives its advantage from the fact that the property of reflection doesn't distinguish between colors, the way lens refraction does. He presented his small telescope to the Royal Society in 1671.

In spite of these improvements, Newton's rather diminutive optical instrument scarcely yielded better results than Galileo's, although it paved the way for larger and more powerful devices. Moreover, the images presented by this rudimentary mirror telescope furthered the great British scientist's empirical understanding of the laws that govern celestial phenomena. It is for these superbly predictive mechanical principles, not his telescope, that he will forever remain renowned.

We previously noted how Newton's principles of motion—which he developed in the 1660s through 1680s, but did not publish until 1687—apply to interacting particles on Earth. Naturally, these principles also pertain to planets and stars. Along with his theory of gravitation, his laws of motion yield a consistent model of the universe, considered entirely accurate until the time of Einstein.

Newton's laws of motion and gravitation helped to continue the secularization of cosmology that began essentially with Galileo. This is in spite of the fact that Newton was a devoutly religious Christian and had no intention to usurp the authority of the Church (in his case, the Anglican Church). But Newton's laws make no explicit reference to God as the prime mover; they are mathematical, not theological, in content. When Newton scanned the sky with his optical device he saw material bodies moving and interacting by means of physical forces, not heavenly orbs circling because of the hand of God.

Newton's cosmos, though not motionless, nevertheless manifests no overall change in appearance over time. In other words, though individual stars and planets exhibit motion through space, they do not evolve into other astronomical bodies, nor do they age. Thus, in a Newtonian framework, the sun, planets, other celestial objects, and space itself are, on the whole, the same for all eternity.

It wasn't until the more powerful telescopes were developed, and the frontiers of observable space were pushed farther and farther back, that scientists began to realize the truly dynamic character of the universe. With these instruments, they started to notice that the planets, stars, and even the structure of space itself all evolve.

The first large reflecting telescope, with a mirror 49 inches in diameter and a focal length of 40 feet, was completed in 1789 by the German-English astronomer William Herschel. This instrument, though little used, represented a significant advance in telescopy; its development was one of the highlights of its creator's impressive career.

One of the greatest astronomers of all time, Herschel, who was born in Hanover, Germany in 1738, originally set out to be a classical musician. Although he developed considerable prowess as an oboist, war cut short his musical career. In 1757, faced with the imminent occupation of his native city by the French army, he fled to England, where he eventually became organist in Bath's Octagon Chapel.

Stargazing, which started out as a hobby, was soon to occupy much of his time. He designed a reflecting telescope and began to use it to observe the heavens. From these detailed observations he made a series of important astronomical discoveries.

In 1773, he studied the motion of 13 stars and discovered that the sun is moving toward the constellation Hercules. This result provided the first clear piece of empirical evidence that our solar system, as a whole, is subject to the same laws of motion that govern the stars and planets individually.

Figure 7. William Herschel, 1738–1822. (Courtesy of AIP Niels Bohr Library, L. Scott Barr Collection.)

Four years later, Herschel decided that he needed a better device from which to perform his measurements. Therefore, he scrapped his original instrument and built a reflecting telescope with a 12-inch-diameter mirror and 20-foot focal length.

Herschel went on to use his technologically advanced instrument to discover the planet Uranus in March 1781. Uranus was revealed to be a planet, not a star, by the fact that it looked like a disk, not a point. The ancients had known only five planets other than Earth—Mercury, Venus, Mars, Jupiter, and Saturn—and had assumed there were no more. With his well-constructed device, Herschel pinpointed the sixth planet, a feat that won him considerable international renown. Because of his discovery, he was elected to the Royal Society of London and appointed by King George III as royal astronomer in residence. Finally he could work full time in the field of astronomy.

Herschel used his new-found time and freedom to conduct a massive survey of the heavens. He developed a theory that the sun is a star belonging to a vast system of stars (now known as the Milky Way galaxy) and proceeded to test this approach by counting the stars in each direction. Ultimately, he wished to determine the galaxy's shape. He published his results in 1785 in a book entitled *On the Construction of the Heavens*, in which he describes the galaxy as a disk with the sun near the center. Although his theory was a phenomenal advance over previous approaches, he made several significant errors. Our galaxy is actually shaped like a spiral with the sun near the periphery, not the middle. Also, Herschel underestimated the size of the Milky Way by three orders of magnitude. Nevertheless, the essential elements of his theory were remarkably correct.

In his later years of stargazing, Herschel became particularly interested in the hazy astronomical objects known as nebulae. Over a period of approximately 20 years, using his 20-foot telescope, he catalogued over 2500 different nebulae. In 1786 he published a volume called the *Catalogue of Nebulae*, which he later expanded into the *New General Catalogue* (NGC), still used by astronomers today.

Herschel found nebulae of several different types. One variety, called planetary nebulae, appeared as dusty clouds surrounding stars. In 1811 he developed a theory, based on these observations, that planetary nebulae eventually collapse into clusters of stars. This theory of stellar evolution was a considerable advance over earlier models in that it no longer assumed that the cosmos was entirely static. Stars were born, aged, and could die. We now know that planetary nebulae are, in fact, the gaseous layers exuded from the outer parts of evolving stars.

Another kind of nebula that he observed seemed to defy explanation. These were the spiral nebulae, which were shaped like twirling pinwheels. Until the 20th century, no one could provide a satisfactory explanation of their origin, composition, and shape. For lack of better explanation, they were assumed to be gaseous clouds similar in size and makeup to planetary nebulae. Furthermore, they were thought to be relatively close to Earth, approximately the same distance as the stars. But we now know that these objects are not clouds of gas at all, but rather galaxies in their own right, millions of light years from our own Milky Way galaxy.

Cosmic Evidence

One of the main differences between religious and scientific approaches to cosmology is their respective takes on the question of whether the truth is malleable. Religion—orthodox religion at least—purports to deliver "eternal" truths that are as relevant today as they were thousands of years ago. For example, consider the case of the Vedic scriptures. Hindus believe that these writings contain truths so timeless that even if the entire universe were to collapse and re-form, the Vedas would reappear in a materially different, but spiritually equivalent, form. A medieval Christian monk would be branded a heretic for suggesting that the Bible could be modified. And the Muslim living in an ultraorthodox culture who doubts the constancy of the Koran's aphorisms would be condemned as a blasphemer and worse.

By contrast, one of the features of scientific truth, as elucidated so brilliantly by Thomas Kuhn (*The Structure of Scientific Revolutions*) and others, is that it is always in flux, metamorphosing itself into a new entity each time a new structural revolution comes along. Therefore, the "heretic" of one generation might be perceived by the next as a prophet.

For example, Euclid's geometry was once considered by mathematicians to be sacred and immutable. When the 19th-century German mathematician Carl Friederich Gauss developed an alternative approach, he was reluctant to publicize it, for fear of appearing unorthodox. Today, non-Euclidean geometry is considered an integral part of mathematics, and the development of this field is viewed as an example of Gauss' foresight and genius. Mathematics, and science in general, are flexible enough to accommodate changing visions of reality.

So it is not surprising that scientific cosmology has taken many different guises in the past century, since our perception of the dimensions and composition of the universe has changed almost continuously. At the time of Herschel, astronomers believed that most of the matter in the cosmos was visible, that it was organized, for the most part, into stars and contained in the Milky Way. Today scientists believe that the universe's material is over 90 percent invisible and that only a minuscule portion is embodied in our own galaxy. How did such a dramatic shift come about?

I'm saving the issue of invisibility for later chapters, because it pertains, in such a significant way, to the question of the fate of the cosmos. But even considering just the *visible* matter, we know today that the vast bulk of it is contained in galaxies other than our own. The Milky Way occupies but a tiny fraction of the known universe; yet in Herschel's day our galaxy was believed to be *the* universe.

The determination of the extent of the cosmos has been a major goal of astronomy for the past hundred years. This quest has been aided vastly by a set of tools developed by astronomers over the years that, in its entirety, is called the *cosmological distance ladder*. This "stairway to the heavens" is a series of distance measurements in which each new result is dependent on the previous ones—like wooden crates precariously stacked to form a stepladder.

It is impossible to overestimate the importance to astrophysics of the cosmological distance ladder. The resolution of the question of the fate of the universe—whether or not it will pass through cycles—is enormously dependent on galactic distance estimates. For without knowing how far away a galaxy is, it is virtually impossible to make numerical predictions as to its long-term behavior. And it is the set of galactic motions that fully determines the large-scale behavior of the universe as a whole.

The survey of distances to Cepheid variable stars, one of the key rungs on the cosmological stepladder, was initiated in 1912 by the astronomer Henrietta Swan Leavitt of the

Harvard College Observatory. Leavitt, a native of Massachusetts, made detailed observations of pulsating Cepheid stars in the region of space known as the Magellanic Clouds. These formations are named for their discovery by the crew of Ferdinand Magellan, who circumnavigated the Earth.

Leavitt noticed an interesting property of these variable stars, one that readily lends itself to accurate distance measurements; they pulsate at a rate that depends on their intrinsic brightness. By measuring the pulsational period of these stars—the time it takes for them to become bright, then dim, then bright again, the amount of light that they produce can be easily calculated. Once this intrinsic luminosity is ascertained, it can be compared to the *measured* brightness of the Cepheid. A simple formula can then be used to find the distance to the star.

The Cepheid variable method can be best understood by analogy. Imagine that, in some particular small town, there were four different types of blinking lights on police emergency vehicles, and each light exhibits one of four different rates of pulsation, and one of four different degrees of brightness. Let's suppose that one lamp uses a 500-watt bulb and pulses once every second; another has a 1000-watt bulb and blinks every two seconds; the third utilizes 1500-watts of power, turning on and off every three seconds; while the fourth measures 2000 watts and flashes every four seconds. In other words, the period of blinking is proportional to the wattage of the bulb.

Now picture the following situation. You are standing at a considerable distance from a police car, just close enough to make out its blinking light. You want to figure out how far away it is. So you begin the "police lamp variable method" by first recording how bright the light appears. Then you measure the periodicity of the lamp, and use the

Expanding Heavens 91

proportionality method to determine its wattage. Finally, you can turn to a chart or formula to determine by what degree brightness tapers off with distance. Typically, the diminution occurs at a rate inversely proportional to the square of the distance. This fact readily yields the distance to the police car. In a similar manner, the apparent brightness and periodicity of a Cepheid variable star can be used to determine how far away it is.

Cepheids are extremely bright (as much as 10,000 times the brightness of the sun) and can be observed as far as 15 million light-years away. Because all galaxies contain these pulsating stars, they provide natural tools with which to figure out intergalactic distances. Leavitt, for example, noticed 17 regularly blinking stars of this type in the Small Magellanic Cloud alone, from which she estimated its distance to be hundreds of thousands of light-years away. This placed the Small and Large Magellanic Clouds at a distance of approximately twice the diameter of the Milky Way—clearly outside our galaxy.

Once the Cepheid variable technique became perfected, in 1912, scientists sought to use it to unravel the mysteries of the spiral nebulae. At the time these distance measurements were being taken, there was considerable controversy over whether the spiral nebulae were galaxies in their own right, or simply unusually shaped denizens of our own galaxy. The former theory, advanced in works such as British astronomer Arthur Eddington's *Stellar Movements and the Structure of the Universe*, suggested that the universe was of a much larger scale than anyone before had dared to imagine. The latter approach—the more conservative one—suggested that the Milky Way contained a larger variety of objects than was previously thought.

As we've seen, it took appropriately grand telescopes to span the immense reaches of space; so naturally a large, state-of-the-art device was needed to solve this puzzle. The clear choice for this task was the newly built Hooker reflecting telescope at the Mount Wilson Observatory in Southern California. This instrument, designed by the great telescope builder George Ellery Hale, utilized a mirror with a 100-inch diameter, at that time the largest telescope reflector in the world.

The young man brought in to supervise this mammoth device, fresh from serving in the American army during World War I, was University of Chicago graduate Edwin Hubble. Hubble began his Mount Wilson stint in 1919 and worked there for a number of years, carefully recording the characteristics of stars and nebulae. He soon became an expert at using the Cepheid variable technique for measuring the distances to remote objects.

In 1924, Hubble aimed the 100-incher at an arm of the spiral nebula classified, according to the Messier catalogue, as M31 (Messier's object 31), and more commonly known as Andromeda. Hubble was fascinated to observe, as he focused the instrument, numerous individual stars forming tiny speckles on the spiral arm. He concluded that Andromeda was a star system in its own right, not just a gaseous cloud belonging to the Milky Way.

To clinch this issue, Hubble identified a dozen Cepheid variables in M31 and measured their frequency of pulsation and their apparent brightness. From these results, he calculated a distance of approximately one million light-years, placing Andromeda far outside the domain of our galaxy. This estimate was later modified (by the inclusion of factors unknown to Hubble) to a distance from us of two million light-years. This is nearly 100 times the distance from our

Figure 8. Walter Adams, James Jeans, and Edwin Hubble at Mt. Wilson Observatory. (From *Images of Einstein Catalogue*. New York: Center for History of Physics, American Institute of Physics, 1979. Courtesy of AIP Niels Bohr Library.)

solar system to the center of the Milky Way. Hubble's findings were the definitive proof that Andromeda—and indeed the bulk of spiral nebulae—represent independent galaxies.

Further measurements of the Andromeda Galaxy have indicated that it contains hundreds of billions of stars and rotates once every few hundred million years. These properties are quite similar to those of the Milky Way; the two galaxies are practically twins. Both are shaped like giant spirals, flattened into disklike structures. (Not all galaxies look like the Milky Way and Andromeda, though; many are shaped like ellipsoids rather than spirals.)

Hubble went on to measure the distances to a number of other so-called nebulae that turned out to be galaxies. By 1929 he had catalogued over 22 galaxies, as far away as the Virgo cluster, using the visible Cepheid variables located in each to calculate how far away it was from Earth. The ease by which he found other galaxies indicated to him that there were a myriad of these star systems in space.

Imagine the shock to the scientific community, and the interested lay public, when it was first revealed that the universe is so big. Not only is Earth but one of nine planets, orbiting a star that is merely one of hundreds of billions in the Milky Way, but, amazingly, our galaxy occupies just a small portion of an unimaginably vast cosmos.

While Hubble was taking distance measurements of these galaxies, he was also performing spectral analyses of each in order to measure its Doppler shift, and hence its speed. Recall that red Doppler shifts indicate motion away from the Milky Way and blueshifts denote motion toward the Milky Way. This phenomenon, called the Doppler effect, is analogous, in sound waves, to the change in pitch of a train whistle as the train heads away from us (lower pitch) or toward us (higher pitch).

Expanding Heavens 95

The Doppler effect was first measured for galaxies in 1912 by Vesto Slipher, an astronomer at the Lowell Observatory. Slipher mapped out the locations of spectral lines for Andromeda and compared what he observed to the standard spectral readings for hydrogen (the most abundant material in the universe). The Doppler shift he found was toward the blue end of the spectrum, indicating that Andromeda was approaching us at a rate of 167 miles per second.

Hubble, in the late 1920s, made a whole catalogue of galactic Doppler shifts and noticed shifts toward both the red and blue ends of the spectrum. However, when he classified these galaxies according to distance, he detected a clear difference between those in the Local Group (nearby galaxies) and those farther away. On the one hand, the Local Group displayed a wide range of behaviors: from blueshifts indicating approach velocities of 210 miles per second, to redshifts signifying recessional speeds of 230 miles per second. This range of speeds and trajectories can be readily explained by the fact that the galaxies in the Local Group are moving under the influence of gravity around their common center of mass.

On the other hand, in the case of more distant galaxies, a radically different sort of behavior was observed, namely, all redshifts. In other words, the Local Group excepted, all other star systems in the cosmos appear to be fleeing from ours. Moreover, Hubble found a simple proportionality between each galaxy's distance and its recessional velocity (now called Hubble's law). These facts seemed to point to the conclusion that the entire universe is expanding.

Interestingly, around the time that Hubble was taking his measurements, the Belgian priest and astronomer Georges Lemaître published an article in a little-known Belgian journal describing his theory of how the cosmos began as a "primeval atom" billions of years ago. This atom exploded and the fiery products of the blast coalesced into the present-day galaxies. This article might have languished in obscurity if it weren't for the fact that Eddington read it and discussed its basic concept with other key astronomers and physicists, including Hubble.

The result of Hubble's findings, Lemaître's ideas and the theoretical work of Einstein and Friedmann (to be discussed) was the development of a comprehensive theory of the origins of the universe, later dubbed the Big Bang model. The key element of this approach—that the recession of the galaxies is a feature of the expansion of space from a point-like entity—has remained an essential part of cosmological models up to the present day.

The Background Hiss of the Great Serpent

Can we be absolutely certain that the universe originated in a Big Bang explosion, as Hubble's evidence appears to indicate? Or might there be other explanations that don't carry such weighty philosophical implications? These were the questions asked by many theorists as they pondered whether it was hasty to assume that the universe was created out of absolutely nothing.

In 1929, soon after Hubble's law was formulated, Swiss astronomer Fritz Zwicky developed a theory of "tired light," an alternative to the notion of an expanding cosmos. In Zwicky's proposal, the cosmological redshift isn't caused by the Doppler effect at all, but rather by the "aging" of photons (the particles that compose light) as they complete their long journey through the universe. The theory states that as these light particles get older they lose a substantial amount of energy and hence shift toward lower frequencies, such as red. The trouble with Zwicky's theory is that there is absolutely no evidence that light particles lose energy in space (except during collisions). Hence, few physicists today assign this notion any credibility.

The most substantial challenge to the Big Bang theory arose via an altogether different line of reasoning. In 1948, three Cambridge astrophysicists, Thomas Gold, Herman Bondi, and Fred Hoyle, advanced the *steady-state theory* of the universe, utilizing a notion called the "perfect cosmological principle" to provide a possible tool for understanding the evolution of the cosmos.

The perfect cosmological principle states simply that the overall appearance of the universe must stay the same for all times. Ten billion years from now, according to this approach, the cosmos would have roughly the same material density and distribution as it has now. This doesn't mean that there can't be changes on a local scale—the death of a star, for instance. But, in the steady-state universe, all changes must even out in the end.

Obviously, the steady-state model proscribes any sudden change in the universe, such as a Big Bang explosion. And that's the whole point; it provides a philosophical refuge for those who don't believe in a beginning of time. It's a third alternative to both cyclical and linear theories of the cosmos—stating that what exists now has always been and will always be.

The way that the steady-state theory explains the cosmological redshift is quite clever. According to this approach, although space expands and galactic spectra are thereby redshifted, matter would be continually and spontaneously created in small amounts to fill in the gaps left by the recession. The amount of material required to fill in these regions would be very small; one atom of hydrogen created every year per 35 million cubic feet of space, would do the job. This matter would form the seeds for new galaxies. Thus, even though the old galaxies would expand outward, threatening to leave large empty regions in space, new galaxies would be produced all the time from the excess hydrogen, enough to fill the gaps and ensure that the universe forever looks the same.

The development of the steady-state alternative to the Big Bang triggered off a war of words between the steady-staters and Big-Bangers that continued for several decades. Leading the fight for the Big Bang approach was George Gamow, a highly respected Russian-born astronomer and author of a number of entertaining popular science books (including the humorous "Mr. Tompkins" series, enjoyed by young scientists for generations). Gamow argued that the cauldronlike conditions of an explosive universal beginning must have been necessary to produce the wide range of elements that we know today. In a research article on the subject of nucleosynthesis (the creation of atomic nuclei), written with his graduate student Ralph Alpher, Gamow showed how the simplest elements, such as deuterium (heavy hydrogen), could have been made in the vast quantities that we presently observe during a fiery Big Bang. He could not, however, offer evidence for his conjecture that *all* of the elements were created that way.

The steady-staters retorted with their own theory that, on the contrary, all of the elements, except hydrogen, were created in stars and spread around the cosmos by supernova explosions. This process would take place in the following manner. First, the hydrogen spontaneously created in space would become absorbed into stars. Then, the fiery cores of these stars would fuse this hydrogen into helium, and he-

Expanding Heavens

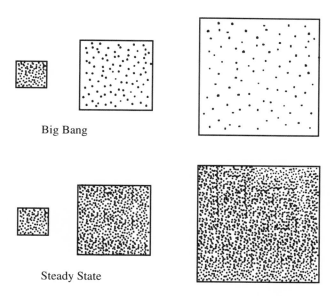

Figure 9. Big Bang vs. steady-state cosmologies. Note that in the Big Bang model, the universe becomes less dense as it expands. By contrast, according to the steady-state approach, new matter is created to fill in the gaps left by the outward movement of older galaxies, resulting, overall, in no change of average density.

lium, in turn, into heavier elements. Gradually, in stars of various ages and types, all of the basic atomic substances that can be produced by fusion—lithium, carbon, oxygen, iron, etc.—would be made this way. Finally, according to this approach, whenever a star exploded in a supernova event, it would seed a sector of the heavens with debris composed of these assorted materials. Thus, after a time a fairly even distribution of all known elements would be achieved.

Interestingly, as it turns out, neither the early Big-Bangers nor the steady-staters were completely correct about the origin of the elements. We now know that all of the deuterium present in the universe today was manufactured during a primordial, billion-degree phase of the universe; stars' cores are simply not hot enough to fuse this simple material from hydrogen. Additionally, it can be shown that the bulk of the helium we currently observe (forming 25 percent of the luminous material in the cosmos), and a substantial portion of the lithium, were produced during an earlier, hotter phase of the cosmos. The measured proportions of hydrogen, deuterium, helium, and lithium seem to match precisely the predictions of the Big Bang theory.

However, it cannot be the case that carbon, oxygen, or any of the heavy elements were fused together from more primitive atoms during a hot-universe phase. By the time that the deuterium, helium, and lithium were produced, the universe had already expanded and cooled down somewhat from its initial fireball state. As the universal density and temperature continued to plummet, conditions rapidly became such that no further nucleosynthesis could take place in space. Only millions of years later, when the first stars were created, could production of the more complex elements take place in their cores. Thus, while the Big Bang theory could explain the coming-to-be of the *simple* elements, it was left to Fred Hoyle and other prominent steady-staters to produce a satisfactory model of the origins of *all* the elements.

Though it became clear by the mid to late 1950s that the Big Bang approach—augmented by the theory of stellar core nucleosynthesis of the heavier elements—was a very successful explanation of the expansion of the universe and the production of its constituents, the steady-staters didn't give up hope. They continued to point out what they thought was a major flaw in the Big Bang theory; while many early

Expanding Heavens

estimates had indicated that the universal explosion occurred two billion years ago, it could readily be shown by geologic dating techniques that the Earth itself is approximately five billion years old. The steady-staters protested that the universe couldn't be younger than the Earth. It wasn't until new estimates of intergalactic distances revealed a much earlier Big Bang that this dilemma was fully resolved. There was no longer a clash between the Big Bang theory's timeline and terrestrial age estimates.

In 1965, the decades-long battle between the Big-Bangers and steady-staters was settled once and for all by a momentous discovery by Bell Lab scientists Arno Penzias and Robert Wilson. Penzias and Wilson were calibrating a large radio antenna, located in Holmdel, New Jersey, in order to convert it for astronomical purposes from use as a satellite signal receiver. They were trying to equip the dish to measure faint radio signals from beyond the plane of our galaxy.

While performing these tests they noticed a curious background hum, no matter in which direction they pointed the receiver. At first they assumed that this "hiss" arose from conventional sources, such as radio interference from nearby cities. They even considered the possibility that pigeon droppings on the dish had somehow caused this strange interference.

After they had ruled out these disparate possibilities, they decided to consult with some of their colleagues. Princeton is not far from Holmdel, and there theorist Robert Dicke was working on a novel test of the predictions of the Big Bang theory. Using an oscillating model of the cosmos with a roughly cyclical time scheme, Dicke calculated that the radiation produced during the universal explosion ought to have filled the universe today with a uniform microwave background of approximately 3 degrees Kelvin. This temperature estimate arose from his assessment of the amount of cooling, due to expansion, undergone by the radiation since the time of the Big Bang. Blackbody radiation (radiation in a closed box—in this case, the universe itself) tends to cool when its container expands at a rate dependent on its initial value.

To understand these results, imagine a balloon filled with hot air, say at 100 degrees Celsius. According to thermodynamics, if one then compressed the balloon, the air temperature would steadily rise as the balloon became smaller. Conversely, if one stretched the balloon by a great amount, the internal temperature would cool down considerably. The same applies to the cosmos as a whole. As it has expanded, the temperature of its radiation has gone down, until it has reached about 3 degrees Kelvin (–270 degrees Celsius).

A Princeton team went out to visit Holmdel, and, sure enough, Dicke's prediction was confirmed. The strange residual noise, uniform in all directions, maintained exactly the right frequency distribution to correspond to the radiation temperature estimated by Dicke. Those viewing these findings had no doubt in their minds that they indeed were perceiving the "background hiss of the cosmos" for the first time in history. Penzias and Wilson had indeed discovered the relic radiation. For these remarkable results, they were awarded the Nobel prize in 1978.

Interestingly, after these results were verified, it was found that Gamow had made a similar prediction years earlier, in the 1940s. Gamow had projected that the radiation produced in the Big Bang would still be present, albeit in a much cooler form. Apparently, none of Gamow's contemporaries had thought to test his premise, so the whole issue of background radiation was completely overlooked until the time that Penzias and Wilson made their discovery.

The evidence collected by the Bell Labs researchers was enough to convince almost all scientists of the reality of the initial cosmic explosion. Today the Big Bang theory has become so widely accepted that it is generally referred to as the *standard model*. The steady-state model, though an integral and respected part of the history of cosmology, is no longer considered by most researchers to be a viable theory of the universe.

FULL OF SOUND AND FURY

> Tomorrow, and tomorrow, and tomorrow,
> Creeps in this petty pace from day to day
> To the last syllable of recorded time,
> And all our yesterdays have lighted fools
> The way to dusty death. Out, out, brief candle!
> Life's but a walking shadow, a poor player
> That struts and frets his hour upon the stage
> And then is heard no more: it is a tale
> Told by an idiot, full of sound and fury,
> Signifying nothing.
> -Shakespeare, Macbeth

An "Ourobouric" Quandary

In recent years a rather clear picture has emerged about the evolution of the universe. This approach, the standard model, has been gleaned from years of experimentation in the fields of elementary particle physics and astrophysics. Thousands and thousands of empirical results obtained in these research areas have been pasted into an ever-growing scrapbook of vital information about the natural world, creating a dynamic journal of an evolving cosmos.

Without a doubt, the combining of these two fields of physics, into an entity often called *astroparticle physics*, has been a catalyst for enormous advances in our quest for a complete representation of the early universe. Astroparticle physics is a strange hybrid of the incomparably large and the extraordinarily minute; it relates the behavior of particles so tiny and near to each other that they can be said to communicate virtually instantaneously, with the interactions of objects so enormous and distant that light signals take millions of years to travel between them. And somehow there is a profound dynamic relationship that bridges the disparate scales of these bodies, spanning the very boundaries of human comprehension. Ultimately, it is hoped that these links will be successfully exploited in the achievement of a deeper understanding of the workings of the cosmos.

There is a certain *ourobouros*-like quality to these interconnections—a particular sort of autosymbiosis involved with first using the large to understand the small and then, in turn, invoking the small to speculate on the behavior of the large. Science writer Marcia Bartusiak reports that "Nobel laureate in physics Sheldon Glashow often likes to draw the ancient symbol of a snake eating its tail to represent this marriage of the microcosm to the macrocosm."[1]

Research ventures during the latter half of the 20th century have yielded a veritable bonanza of information about the microcosmic realm of elementary particles. Atom smashers of all sorts—cyclotrons, linear accelerators, nuclear reactors—have revealed that all of the detectable matter and energy in the world is composed of two sorts of particles, fermions and bosons, categorized by a structural property called spin.

Spin is a quantum theoretical classification introduced in 1925 to account for particular discrepancies that were found in atomic spectral lines. In examining the properties of atoms, physicists recorded extra spectral lines that couldn't be explained by what was then known about quantum mechanics. Theorists found, however, that they could account for these aberrant lines by assuming that electrons behaved like spinning spheres of charge. To characterize this spinning behavior, they incorporated the property called spin into quantum theory.

Physicists later discovered that all elementary particles can be classified by their spin. Spin quantities are determined by placing particles in electromagnetic fields and noting the resulting behavior. Particles with half-integer spin are called fermions, and those with zero or integer spin are called bosons.

Fermions, which are divided, in turn, into quarks and leptons, form the building blocks of all known material objects. (While the protons and neutrons, the constituents of atomic nuclei, are composed of three quarks each, electrons, the leptons comprising the periphery of atoms, are considered indivisible.) There are six flavors (kinds) of quarks: up, down, strange, charmed, top and bottom. Types of leptons include electrons, positrons (electrons' positively charged counterparts), neutrinos (seemingly massless leptons), and antineutrinos, as well a host of other elementary particles.

Bosons provide the glue that holds matter together, as well as the repulsive energy that often tears it apart. Photons, gravitons (carriers of the gravitational force), W and Z bosons (conveyers of the weak force that mediates radioactive decay), and gluons (agents of the strong force that holds nuclei together) are all good examples of this second major category of particles.

Traditionally, it has been expected that accelerators would provide us with a host of new information about the particle zoo. For years it has been hoped, for instance, that powerful atom smashers would help us to understand why two sorts of material constituents, fermions and bosons, exist, and would aid us in determining whether there once was only one kind of particle, as some theorists surmise.

Unfortunately, each new type of accelerator built must necessarily be bigger, require much more energy, and hence be far more expensive than its predecessor. These cruel facts make it increasingly harder to justify the cost of these machines, especially in an age of tight budgets and widespread social problems.

As a result of these and money-saving, but I believe shortsighted, considerations, the controversial Superconducting Super Collider (SSC), a giant, state-of-the-art accelerator that was scheduled to be built in Texas, was canceled in 1993 during its early construction phase by the U.S. Congress. If we assume that funds for this project are never restored—a safe assumption—it is clear that elementary particle research has been dealt a heavy blow.

Astroparticle physics, however, provides us with a way out of this dilemma by turning the early universe into a giant particle laboratory. By making astrophysical measurements of the conditions at the beginning of the observable universe, one can construct sophisticated models of the particle interactions that may have caused such conditions. Thus, if one becomes fully convinced, based on cosmological arguments, that a given particle or interaction must exist, astroparticle physics reduces the need for expensive accelerators to produce the same results (though, if at all possible, one should always double-check results by using several independent methods).

With all these references to the *early* universe and the *initial* cosmic explosion, one might wonder if we aren't making the implicit assumption that the Big Bang was indeed the absolute beginning of time. What about the notion of cyclical creation and destruction of the cosmos? Perhaps the Big Bang was not a true beginning at all, but simply one particular episode in cosmic history, preceded—and destined to be succeeded—by numerous other all-encompassing explosions.

Let's put these questions aside for the moment. Although the issue of what preceded the Big Bang is a deep and important question in cosmology, many astroparticle physicists prefer to start the count at several milliseconds *after* the explosion and call this period the early universe, for want of a better mode of description.

The First Few Seconds

Scientists today have a reasonably clear picture of what the cosmos was like approximately the first hundredth of a second after the Big Bang. At this time the density of the universe was several billion times that of water. Its temperature was also immense—over a hundred billion degrees. By comparison, the temperature at which lead boils is only about 2000 degrees Kelvin.

Given the Hubble expansion (enlargement of space) since that time, the universe back then was far more compact than today. Because the present-day size of the cosmos is a matter of contention (it could be that the universe is infinite), we really don't know its absolute size during that earlier era. We can only estimate its relative dimensions; because cosmic expansion is inversely proportional to cosmic temperature decrease, we know that its distance scale was some 30 billion times smaller than at present.

Furthermore, when the universe was only a hundredth of a second old, its composition was much simpler than now. Instead of a complex array of atoms and molecules of various masses, sizes, and constituents, the cosmos back then consisted of a sort of elementary particle "soup." Like an alphabet stock, chosen to have equal amounts of each letter, this cosmic soup contained proportional quantities of electrons, positrons, neutrinos, antineutrinos and photons. In contrast to the wide range of speeds and interaction distances particles now hold (the neutrino, for instance, can today travel millions of miles without affecting or being affected by another particle), at the hundredth of a second mark, all five of these particles, being in thermal equilibrium, acted quite similarly. They consistently underwent innumerable rapid-fire collisions each second; the mean free paths (intervals of freedom between collisions) were exceedingly brief. This was true for electron motion, light radiation and neutrino radiation as well. Thus, the dynamics of the universe was far more uniform than at present.

This sizzling broth of particles was not precisely homogeneous, however. Mixed in with the evenly distributed electrons, positrons, neutrinos, and photons were a meager number of larger "chunks," adding variety to the concoction. These heavier particles, the nucleons (neutrons and protons in equal numbers), were present in a ratio of about one per every *billion* electrons, neutrinos, or photons. Eventually these nucleons would form the core of visible matter in the universe; in this ultrahot period, though, they barely played a role.

Finally, there was, in all likelihood, another major constituent of the early cosmos, the so-called dark matter. Physicists theorize that as much as 99 percent of the matter in the cosmos doesn't produce detectable radiation. There are a number of reasons for this belief, mainly having to do with the unexpectedly large orbital velocities of stars occupying the galactic periphery. Suggestions that this discrepancy is caused by the gravitational effects of invisible matter have led scientists to a extensive search for this missing material. And many of the dark matter theories involve massive particles that were present, alongside the other elementary constituents, during the hot early universe.

As the universe expanded and cooled down, a number of important changes took place in its composition. First of all, during the first few seconds of the universe, the overwhelming majority of electrons and positrons found it energetically favorable to annihilate each other and turn into photons (pure light energy, that is). One pair of photons was created for every electron–positron couple destroyed. Thus, the quantity of light energy grew enormously during this interval, considerably overtaking the amount of neutrino radiation present in the universe.

Second, as the temperature fell, conditions became more and more favorable for neutrons to turn into protons by emitting electrons and neutrinos (because, as particle theory tells us, the conversion process for neutrons, called *beta decay*, is highly dependent on temperature). By the time the temperature of the cosmos dropped below a billion degrees Kelvin, the ratio of protons to neutrons had reached its ultimate value of slightly less than seven to one.

At this point, approximately three minutes after the Big Bang, a fortuitous event occurred that made possible the existence of life on Earth today. Just as the remaining neutrons started to undergo decay, en masse, into protons, the temperature of the universe gradually became cool enough for elementary nuclear fusion to occur. Almost all of the free neutrons began to join together with protons to form the nuclei of deuterium atoms. Bound neutrons tend to be stable, so the conversion process virtually ceased after this fusion occurred. Thus, from this point forward, the ratio of protons to neutrons became locked into place.

We are rather lucky that so much deuterium was formed during this particular early stage of the universe. Deuterium is a difficult element to produce; if it is too hot, it flies apart immediately. Therefore, the universe had to cool down for a few minutes before such production was possible. If, however, deuterium creation required even *lower* temperatures, not enough neutrons would be around to produce much of it. Most of them would have already decayed away. It's fortunate that the binding energy of deuterium is the value that it is; otherwise, deuterium might never have been formed.

And if deuterium were never created you wouldn't be reading this passage. All other elements besides hydrogen, including those essential for life, are built up from deuterium in the process of nucleosynthesis. These substances are far more stable than deuterium. Hence, many astrophysicists refer to the tricky juncture that had to be passed in the early cosmos as the "deuterium bottleneck."

Once a sufficient reservoir of deuterium was formed, a number of other elements were built up very quickly. Each time a deuterium nucleus collided with a proton, a light isotope, called helium-3, was formed. When deuterium interacted with a neutron, tritium, the heaviest form of hydrogen, was created. Helium-3 and tritium collided, in turn, with neutrons and protons to form the abundant isotope helium-4.

Finally, a small quantity of helium was converted into lithium by a similar mechanism. This process continued until the universe expanded so much and became so materially sparse that no further reactions could take place. Hence, all of the heavy elements produced in the cosmos thereafter needed to be forged millions of years later in the molten cores of stars. Thus ended the age of Big Bang nucleosynthesis.

Today we can readily measure the amount of hydrogen, deuterium, helium, and lithium produced by the Big Bang. Through spectral analysis and other techniques, a precise determination can be made of the proportions of these substances in space. In outstanding confirmation of the standard model of cosmology, these ratios correspond exceeedingly well to theoretical predictions.

The Uniform Sky

As a city dweller, surrounded by the perpetual hazy glow of street lamps, I've rarely had the experience of seeing many stars in the sky. Several years ago, however, I had the opportunity to visit the Franz Josef glacier in a remote part of the South Island of New Zealand. Walking along a deserted country road on a moonless night, I was astonished to witness the *nearness* and *fullness* of the stellar canopy above me. Aside from celestial features easy to identify, such as the Southern Cross and the Milky Way, the sky looked remarkably consistent, almost as if a mammoth spray can had dispersed white speckles uniformly across the heavenly dome.

Is the cosmos truly homogeneous? That is, does the sky appear the same from every vantage point in space? Unfortunately, our space probes haven't traveled far enough away from Earth to answer this question directly.

Humility, though, provides us with a straightforward, indirect way of determining the homogeneity of the uni-

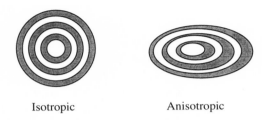

Isotropic   Anisotropic

Figure 10. Isotropic vs. anisotropic geometries.

verse. Scientists modestly assume that the Earth occupies no special place in the cosmic domain, that whatever is true for our world must be true for other worlds as well. This is called the *Cosmological Principle*.

Any measurements we take of the uniformity of our own sky, a property called *spatial isotropy*, provide a telling indication of the homogeneity of the universe. For if the heavens look roughly the same in all directions from our own vantage point, they should look basically the same in all directions from every location in space. As cosmologists generally put this, spatial isotropy implies spatial homogeneity.

Galactic surveys and material distribution analyses indicate that the sky is isotropic as far as can be gauged. The number of galaxies per unit solid angle, and the proportion of hydrogen to helium and to other elements, appears to be the same no matter to which direction of the sky one turns.

Perhaps the most significant indicator of spatial isotropy is the cosmic background radiation distribution. Since the discovery of radiation by Penzias and Wilson, scientists have sought to determine whether or not any ripples might be found in its seeming spatial uniformity. For the first decade after the Bell Lab findings were announced, however, not a single researcher could detect the slightest glitch in the microwave background's uniform temperature of about 2.73 degrees Kelvin. Because background radiation provides a revealing portrayal of the matter distribution of the early universe, astrophysicists remained convinced of cosmic homogeneity.

In 1977, a team led by George Smoot of Lawrence Berkeley Laboratory and the Space Sciences Lab of the University of California at Berkeley announced results that provided further evidence for a homogeneous universe. Obtaining its data from high up in a NASA Ames U-2 jet (to avoid interference from terrestrial signals), they reported the discovery of a "dipole anisotropy" in the cosmic microwave background. Basically, they noticed slight temperature variations when they pointed their instruments at opposite directions in the microwave sky.

There was no contradiction, however, between the Smoot team's results and the notion that the microwave background is essentially isotropic. The dipole anisotropy came from a noncosmological source: the motion of our galaxy relative to deep space. As our galaxy travels through the cosmos, its velocity relative to the microwave background causes a distinct Doppler effect. Because of this Doppler shifting, the sky appears hotter in the direction that our galaxy is heading, and cooler in the backward direction. Hence, the dipole anisotropy is not a genuine property of the background radiation, but rather an illusion caused by our own motion.

Once this dipole effect was subtracted from their results, Smoot and his colleagues found that the cosmic background was isotropic to an outstanding degree; its temperature varied less than 1 part in 10 thousand.

But Smoot was determined to push further; he wanted to know for himself if there were any anisotropies at all in the cosmos.

It was not until 1992 that real anisotropies, independent of the Milky Way's motion, were found in the cosmic microwave background. With the use of special instruments, called differential microwave radiometers (DMR), designed by Smoot and launched aboard NASA's Cosmic Background Explorer (COBE) satellite, it was found that there were indeed minute anisotropies in the Big Bang relic radiation—a spread of roughly six parts per *million*.

These "ripples" are most likely remnants of energy fluctuations due to the primordial seeds that formed the progenitors of galaxies in the very early universe.

Still, even taking into account these much heralded COBE findings, the cosmic microwave background is incredibly uniform. Imagine a white canvas, measuring one square meter, speckled with only a few square millimeters of black paint. That's how much these ripples in the microwave background stand out.

Further measurements have confirmed the COBE results—that the background temperature is approximately 2.73 degrees Kelvin everywhere, with near uniformity. In 1993, Katherine Roth and David Meyer of Northwestern University, working with Isabel Hawkins of the University of California at Berkeley, measured the temperature of interstellar clouds by studying the microwave radio emissions of a carbon–nitrogen compound, called cyanogen, present in these clouds.

Cyanogen has the property of absorbing energy from the cosmic background and reemitting it at a characteristic frequency. Thus, assuming that these clouds are in thermal equilibrium (at the same temperature) as the background radiation, the temperature of the latter can be readily measured.

The results of Roth, Meyer, and Hawkins turned out to be in close agreement with Smoot's findings. These studies clearly ruled out a local origin for the microwave radiation and further helped confirm the Big Bang model's prediction of a universal thermal background.

Crossing Horizons

Why is the cosmic microwave background so uniform? One might think that the radiation has the same temperature because it originated from a single source. This question becomes far more profound, however, when one realizes that the microwave radiation came from myriad different sources, physically separated by enormous distances. This is because this radiation was set loose into the cosmos not during the first moments of the Big Bang, but rather during a time about a hundred thousand years later.

The period in which the photons that form the cosmic background were generated is called the *era of recombination*. This epoch is characterized by a decoupling of radiation from matter during the formation of neutral atoms from the hot universal plasma (made up of negatively charged electrons as well as positively charged hydrogen, helium, and lithium ions). Unlike charged ions, neutral atoms do not couple with photons and therefore roam freely. In other words, after each ion in the plasma grabbed enough electrons to neutralize itself, a massive amount of radiation was released, and it is this burst of energy, albeit cooled down considerably, that we see evidence of today.

Thus, the issue of why the microwave background radiation is almost completely uniform in temperature is strongly related to the question of how the spatially dispersed atoms created during the era of recombination coordinated their temperatures so well. For if the atoms formed in the plasma weren't at the same temperature, the photons released also wouldn't be at the same temperature. But these photons were; so, how did the distinct parts of the plasma "know" that they should have the same temperature?

Perhaps these distant atoms somehow communicated with each other to equalize their amounts of energy. By "communicate," I'm not implying an act of intelligence. I simply mean that they sent signals from one to another. If this sort of exchange were possible, it would explain why thermal equilibrium existed during recombination. The process would be analogous to the heat exchange that turns a glass of hot tea into cold tea (of uniform temperature) with the addition of an ice cube.

Unfortunately, the standard Big Bang model presents us with an obstacle in resolving this dilemma, known as the "horizon problem." There simply isn't enough time between the beginning of the universe and the era of recombination for light signals to travel from one end of the universe to the other and equalize all the disparate temperatures.

The maximum distance that a light pulse could have traveled since the Big Bang is called the *horizon length.* If, on the one hand, two particles are within one horizon length of each other, then it is possible that they have communicated once; one could have sent a light pulse that the other received. On the other hand, if two objects lie a distance greater than one horizon length from each other, then the possibility that they have exchanged a signal must be excluded.

According to the Big Bang scenario, it can be estimated that, during the era of recombination, the distance across the universe was over 90 horizon lengths. It would have been impossible for the hot plasma to have coordinated itself in temperature to any large degree. Yet it did—within one part in a hundred thousand across the entire cosmos.

The Age of Inflation

The inflationary universe scenario was developed in the 1980s as a way of addressing the horizon problem and several other technical difficulties with the standard Big Bang model. The basic idea behind the original picture of inflation is that the cosmos went through a period of extremely rapid expansion during the very early universe—some 10^{-35} seconds after the Big Bang. Then, somehow, this inflationary period ended, and the standard Hubble expansion that we observe today began.

The assumption of an inflationary epoch provides a straightforward way of eliminating the horizon problem. To follow this line of reasoning, we first consider the point in which the universe was extraordinarily small (approximately a trillionth of a trillionth of an inch across). Now we suppose, in contrast to the standard Big Bang theory, which assumes monotonic growth, that the universe went through a period of exponential expansion. During the tiniest fraction of a second, all of space literally blew up from about a billionth the size of an electron to roughly the size of a grapefruit.

To picture how fast the universe expanded under inflation, imagine consuming a miracle hair growth solution that works in an extremely brief amount of time. You wish to test this elixir and begin to take a sip. As soon as the concoction has passed through your lips, your hair has grown from a few inches to billions of billions of miles in length, rapidly surpassing, in short order, the physical dimensions of the galaxy. Such was, according to theory, the explosive growth of the cosmos during the inflationary era.

It is easy to understand how the horizon problem disappears. The entire observed universe, all the galaxies that we see around us, has evolved from a region of space over a trillion trillion times smaller. Let's call this region the "protocosmos." Before inflation, the protocosmos was so tiny that all of its parts were in complete communication. Therefore, it possessed an even density and temperature throughout.

Perhaps, in this interval before inflation, there were other regions of space besides the protocosmos. These other sectors may have possessed radically different temperatures and densities. However, these disparate regions could never be observed today, since their light could never have reached us. Long ago, inflation would have propeled these domains far beyond the current range of observability. Thus, any part of the sky from which we presently receive signals has emerged, by definition, from the protocosmos, not from any larger region.

It is clear, according to this scenario, why the observable universe is so homogenous; it is a blown-up portion of a small, causally connected, uniform region of space, that is, an area of dimensions less than one horizon length. Any inhomogeneous regions, therefore, lie beyond observability; they have literally been blown away.

In our hair tonic analogy, one can readily see how "hair color homogeneity" might be produced by inflation. Let us assume for the sake of argument that your hair is naturally bright red. Suppose you put a bit of green dye on a small piece of your hair before consuming the miracle growth formula. You also put other colors—blue, aquamarine, in-

Sound and Fury

digo—on other strands of your hair. Your hair is now clearly inhomogeneous.

Now you gulp down a bit of Miracle Growth. Within nanoseconds your hair has crossed the Pleiades. You look up at a strand and there is bright red as far as the eye can see. All the other colors now lie on parts of your hair forever beyond your visual field. For all intents and purposes, the observable region of your hair is now homogeneous in color. Inflation has produced effective uniformity.

There are several different sorts of inflationary models, each speculating on what may have caused the era of rapid expansion and what may have ended it. An early version of inflation, proposed in 1979 by Alexei Starobinsky of the Lev D. Landau Institute in Moscow, was popular in Russia but never caught on internationally. The most prominent early approach, now called "old inflation," was proposed in 1980 by Alan Guth (who was at Stanford at the time and is now at MIT). Guth drew from the areas of elementary particle physics and quantum field theory to postulate a way in which the early cosmos blew up suddenly.

Guth came late to the field of cosmology, but though his entrance was belated, it was not without notice. Upon completing a Ph.D. from the Massachusetts Institute of Technology, he thought that the science of studying the universe was too abstract a discipline for him. Instead, he sought to understand the world of particles by using powerful mathematical techniques to simplify the complexities of high-energy interactions. While at Stanford, which he attended after a few postdoctoral stints and an instructorship, he worked on developing means of representing the known forces of nature by mathematical models called "gauge theories."

Guth's studies at Stanford centered around an area of field theory known as grand unification, an idea with a long history. Until the late 19th century, it was universally believed that electricity and magnetism were two different phenomena. However, James Clerk Maxwell showed that they could both be incorporated into a unified electromagnetic theory. About a century later, Stephen Weinberg and Abdus Salam proposed a theory that combines the weak and electromagnetic forces. In 1983, this theory was confirmed when the electroweak intermediators, the W and Z bosons, were discovered by two experimental groups working at the CERN accelerator in Geneva.

Guth and other researchers were trying to continue this tradition by incorporating the strong force, along with the electroweak, into an integrated approach, called a Grand Unified Theory (GUT). Their hope was that any theory developed might be verified by massive particle accelerators, such as the proposed SSC. Some progress was made, but unfortunately no complete theory could be devised and tested. Guth was not discouraged, however, and continued to work hard on GUT.

A successful GUT would model three of the forces of nature, the strong, weak, and electromagnetic—as a single entity. The fourth force, gravity, is excluded from GUT and only considered in even more esoteric, ultra-high energy approaches known as quantum gravity theories. Because these forces act in nature quite distinctly, with a wide range of strengths and interaction lengths, it is a challenge to imagine them behaving as one unit.

The strong force, which is associated with nuclear energy, acts only on hadrons: the particles built up of quarks, such as neutrons, protons, and mesons (bosonic hadrons). It is significant only for elementary particle interactions on a minuscule scale, never on a larger scale. The weak force, which is responsible for beta decay, also operates on a tiny scale. In contrast, though, to the strong force, hadrons and nonhadrons (electrons and neutrinos, for example) experience the weak interaction.

The third force, electromagnetism, operates on both small and large scales, from the subatomic to the planetary. Only charged objects experience this interaction; neutrons, for instance, are excluded. Thus, with electromagnetism, the strong and weak forces all normally operate in quite disparate arenas.

According to GUT, however, at high enough temperatures, all of these forces should behave identically. At temperatures in the billions of billions of degrees, there should be but a single interaction, operating at a common range. A successful Grand Unified Theory would be the unitary approach that physicists have sought for decades.

In 1978, Guth attended a lecture that would change his career path significantly by turning his gaze from the minuscule to the all-encompassingly large. This talk, by Robert Dicke, addressed some of the paradoxes of the Big Bang theory, and detailed the history of trying to resolve these issues. Guth walked away from the lecture eager to apply his knowledge of particle physics to help unravel some of the mysteries of the cosmos. He hoped, as a premium, that he could use cosmology to test out some of his GUT ideas.

About the time of the New Year in 1980, it occurred to Guth that a brief inflationary era would smooth out many of the seemingly intractable difficulties of the standard cosmological approach, such as the horizon problem. And, moreover, he realized that one of the many predictions of GUT could provide the mechanism for inflation to occur.

In the standard GUT, there is a phase transition between two universal vacuums, called the "false" and "true" vacuums. In theoretical physics, the vacuum is the lowest energy state available, the state in which a decaying particle will eventually find itself after losing its energy. But this bedrock state can change, as temperature is lowered, in a phase transition analogous to the transformation of water to ice. When the temperature of liquid water is lowered, its molecular constituents begin to find themselves pining for more energetically favorable configurations. As the temperature reaches 32 degrees Fahrenheit, portions of the fluid start to coagulate into ice. Eventually, as more and more heat is removed, a block of ice is formed. This is the common phase transition for water, known as freezing.

Interestingly, if one is extremely careful and uses water with few impurities, one can put off this transition to a lower temperature in a process known as *supercooling*. Supercooled water is liquid water that has been slowly cooled in a way that allows it to remain fluid far below 32 degrees. Although it has the same temperature as ordinary ice, it remains in an alternative higher energy state. The extra energy retained by the supercooled fluid is called the *latent heat*, and it is released if the water is further cooled and freezes. Once the supercooled water turns to ice, its latent heat is lost as it arranges itself in a more energetically favorable situation.

Guth noted that an analogous process to supercooling can take place in a GUT phase transition between a false and true vacuum. This result is based on a model developed in the 1970s by Russian physicists David Kirzhnits and Andrei Linde. If one lowers the temperature of a symmetric arrangement of forces—with all forces having equal interaction strengths and ranges—a transformation takes place to a configuration in which the symmetry is broken, and there are differences between the forces (that is, the strong force and electroweak force become separate entities). Since there is a tendency for objects or structures to move or otherwise transform themselves in a way that lowers their total potential energy (a ball releasing potential energy by rolling down a hill is a good example of this), the original symmetric configuration expels energy as it spontaneously decays into the final state of broken symmetry. Thus, the end result is that the false vacuum, named because it doesn't represent the absolute lowest energy state but, rather, a transitional energy state, has spontaneously transformed itself into true vacuum.

But there is a caveat to all this. As in the case of supercooled water, it isn't necessary for all of the false vacuum to decay into true vacuum at once. Pieces of the false vacuum might remain, trapped in the real vacuum-like ice chunks floating in a cold stream. Or perhaps even most of the false vacuum persists, with only "bubbles" of true vacuum forming here and there.

Now apply this to the cosmos. Suppose that the universe during the GUT transition remained, for quite some time, in the false vacuum. "Quite some time," is, of course, relative. Remember that the universe was only 10^{-35} seconds old at the beginning of this era. Now the false vacuum, being more energetic than the true vacuum, possessed a fixed, nonzero energy density. And, as we'll see, matter and energy are the engines that have driven the universe to expand. Therefore, the energy stored in the pretransition configuration served to propel the universe outward explosively. One can mathematically show, as Guth did, that an exponential expansion ensued, which caused the universe to inflate by more than a factor of 10^{25}.

However, as Guth and other researchers soon realized, there are a number of difficulties with old inflation. The main problem, called the "graceful exit dilemma," concerns the means by which the inflationary period ends and the conventional Hubble expansion begins. Guth found that the true vacuum bubbles, formed amidst the false vacuum background, could not expand fast enough to fill the whole universe. Moreover, if these bubbles happened to collide with each other, they would make the universe extremely inhomogeneous.

To address this predicament, Andrei Linde, working in Moscow, introduced in 1982 a model known as "new inflation." Andreas Albrecht and Paul Steinhardt of the University of Pennsylvania independently developed a similar model later the same year. In this approach, the graceful exit dilemma is eliminated by assuming that the observable universe has evolved from a single bubble, rather than from many bubbles.

But new inflation was found to have problems of its own. It required one to assume a highly specialized set of initial conditions for the cosmos—enough to produce the proper amount of supercooling. Rather than inflation occurring naturally, it needed to be "fine-tuned" into being. This stipulation clashed with a generally held notion that a good cosmological model should be one that holds true for a wide range of parameters.

In 1983, Linde discovered a series of results that led him to abandon new inflation and propose the "chaotic inflation" model. Unlike previous approaches, chaotic inflation has the advantage of not requiring supercooling. Therefore, as Linde showed, it does not depend on the fine tuning of parameters for any particular elementary particle theory. Nor does it include the standard assumption that the universe was hot from the very beginning. All it requires is the existence of scalar fields (physical entities with special theoretical properties) during the earliest stages of the universe.

Sound and Fury

Linde explained to me in a phone conversation that his decision to abandon the notion of supercooling was very painful, because it seemed such an elegant and simple concept. As he put it, "the idea of supercooling was so seductive that it was very difficult to move on to a better model."[2]

Nevertheless, once cosmologists began to realize the severe limitations of new inflation, the idea of chaotic inflation became increasingly attractive. Linde explained why he came to believe that the observable universe was produced during an era of chaotic inflation.

> In chaotic inflation one considers different parts of the universe (or different 'universes'), containing different scalar fields. Those parts of the universe where the scalar field is small will not inflate and will remain small forever, or will rapidly collapse. We cannot live in such domains, so they are not interesting for us. On the other hand, those parts where the scalar field is large will expand exponentially, and give rise to the conditions which are necessary for existence of life. Thus, inflation could create order out of chaos, and the total volume of all parts of the universe where inflation has occurred is much greater than the total volume of those 'losers' where it did not happen.[3]

In other words, our part of the universe is the result of the exponential expansion of a region of space with a sufficiently large scalar field. We know that we live in an expanded part of space, because life can evolve only in sectors of the universe large enough to contain galaxies. Other regions, with smaller scalar fields, never expanded. Therefore they do not contain life and have been pushed, by the expansion of our own sector, beyond the range of present-day observability.

Galactic Seeds

Inflation has the effect of smoothing out all inhomogeneities by essentially casting them out of sight: "out of sight, out of mind," goes the cliché. By the time the universe emerged from the inflationary regime and entered into its slower Hubble expansion, it should have been as uniform as one could imagine—so uniform that there shouldn't be inhomogeneities today on any scale.

Then how can one explain the presence of medium-scale inhomogeneities in the present-day sky: galaxies, galactic clusters (collections of galaxies) and superclusters (sets of clusters)? Though the universe on the *largest* scale seems to have an even distribution, on the *medium* scale it is impossible to ignore the existence of complex structures. The fact is, some regions of the cosmos, such as the so-called voids, have relatively few stars, while others, such as the centers of galaxies, are literally packed with stars.

Physicists have grappled with the problem of structure formation for decades. It was a difficult task to explain how galactic-sized clumps were formed in a universe that, as a whole, expanded equally outward in all directions. Gravity was undoubtedly the agent involved in producing these structures, but many questions remain. Why are there voids in the sky containing few galaxies and other areas with many? Why did gravity select, in particular, structures of galactic, cluster and supercluster sizes and not other sizes? How, in a homogeneous cosmos, did seeds of inhomogeneity form? At what stage in cosmic history did the lumpiness occur? And why is there little evidence today of the processes leading to galactic formation?

Astrophysicists today are relying on COBE to answer many, if not all, of these vital questions. Anticipation of this experiment has been building for years. Until COBE began beaming back to Earth a portrait of the low-temperature radio sky, there was no signal in the microwave background to indicate the presence of inhomogeneities. Scientists were

Sound and Fury

hard pressed to explain where and when the "clumps" that evolved into galaxies arose.

The launching of the COBE satellite on November 18, 1989, with its three pinpoint-accuracy radiometers on board, was a tremendous cause for rejoicing among radio astronomers. With its orbit some 900 kilometers above Earth, it is in a perfect position to measure signals from deep space; it is high enough that the Earth's atmospheric effects do not bias its instruments yet low enough to avoid too much interference from the charged particles encircling the Earth. Furthermore, to increase its accuracy even more, sophisticated computers on Earth monitor its signals and subtract undesirable effects not related to cosmic background radiation. Overall, the COBE satellite, including its differential microwave radiometers, is the most precise instrument of its kind ever invented, giving scientists enormous hope in using it to help unravel the mysteries of the early universe.

On April 23, 1992, George Smoot, as principle investigator of the COBE team, announced to the press the discovery of ripples in the Big Bang microwave radiation. These ripples manifested themselves as "hot" and "cold" portions of the sky, with temperature differences of less than a hundred-thousandth of a degree.

Quite a stir was caused by this announcement of the first detailed map of the Big Bang. Newspaper headlines proclaimed it the "discovery of the century, perhaps of all time," and the "holy grail of cosmology." More sober assessments by science writers and professors generally ranked it the third most significant find of the century in cosmology, next to Hubble's discovery of the expansion of the universe and Penzias and Wilson's detection of the cosmic microwave background radiation. Around the world it was recognized that the Smoot team's observation of these ripples heralded a new era in physical cosmology.

The manifest explanation for these signal variations in the cosmic microwave background is that when this radiation was produced, billions of years ago, the cosmos contained small density variations throughout, reflected in an unequal temperature distribution. These density inhomogeneities were the seeds that pulled matter together, under gravity, to form the progenitors of galaxies. The ripples in temperature that Smoot's team observed are veritable relics of the period of structure formation in the universe. The information obtained from the scale and distribution of these irregularities will serve as a litmus test for all cosmological theories to come.

Proponents of the inflationary universe model, in all of its incarnations, are especially buoyed by this discovery. In recent years, theorists have developed a detailed picture of structure formation under inflation, based on the production and subsequent dilation of small fluctuations in the false vacuum. These "fluctuations," theorists argue, arise from the Heisenberg uncertainty principle's stipulation that no vacuum can be free of particles and fields for all times; therefore, irregularities continually arise and decay. This approach led to a prediction that any ripples in the microwave background should have a scale-invariant structure; that is, they should appear the same on all scales of measurement. Remarkably, to a large extent, the COBE results have borne this out.

Scientists continue to look for pieces in the puzzle that was the very early universe. We come closer and closer each year to a full understanding of the first few moments of the Big Bang. But what of the time before the Big Bang? What produced this universal outpouring, or did it somehow

Sound and Fury

Figure 11. Paul Steinhardt, b. 1952. (Courtesy of Paul Steinhardt.)

produce itself? And what will become of the universe in its twilight years? Will it fade into quiescence or begin a catastrophic collapse and reexpansion?

These are questions that are extraordinarily difficult for cosmologists to answer. They concern areas that were once reserved for metaphysicists and theologians. Increasingly, however, scientists turn to the powerful techniques single-handedly developed by Einstein over three-quarters of a century ago, and try to settle these powerful issues. Even if some of these mysteries are never resolved, scientists hope, at the very least, to wring from nature all of the information that it will yield about our universal fate.

Figure 12. George Smoot. (Courtesy of George Smoot and Lawrence Berkeley Laboratory.)

In the next chapter, we'll begin a detailed look at what astronomers and physicists, using the tremendous potency of Einstein's equations and the precise tools of observational cosmology, have determined so far about the ultimate fate of the cosmos.

PART 3

THE QUEST FOR OUR UNIVERSAL DESTINY

MAPPING OUR FATE

Space acts on matter, telling it how to move. In turn, matter reacts back on space, telling it how to curve.
-*Misner, Thorne, and Wheeler, Gravitation*

Singular Genius

In ancient times it was thought that gifted men and women—oracles, magicians, prophets, and priests—could predict the future, even forecast when the world might end, or if time might continue, in cycles, forever. Modern science no longer relies on tea leaves and tarot cards to divine the destiny of the cosmos. Yet we still look to the skies to foretell our fate, and we still believe that the rare man can produce, from his instincts and genius, a blueprint for our universal prospects.

General relativity, the physical theory that predicts the behavior of space by knowledge of its contents, is one of the few areas of science that arose from a single man's prophetic vision.

Although numerous scientists have analyzed relativity and have obtained results from its fundamental equations, the only man to devise a theory whose basics have never

Figure 13. Albert Einstein, 1879–1955. (From *Images of Einstein Catalogue*. New York: Center for History of Physics, American Institute of Physics, 1979. Courtesy of AIP Niels Bohr Library.)

been altered was Albert Einstein. And it is through his magnum opus that we hope to know someday if the universe will last forever—repeating itself in an endless chain of cycles—or if it is doomed to die once and for all.

Einstein was born on March 14, 1879 to a Jewish family in Ulm, a medium-sized town in the Swabian region of

Germany. He obtained his primary schooling in Munich and the bulk of his secondary education at the Zurich Institute of Technology in Switzerland. Although he was bright, he obtained little recognition or support for his work. He settled, after graduation, for an unsatisfying job as a high school teacher, before obtaining a position as a Swiss Patent Officer.

Einstein was fascinated by the workings of the universe, and couldn't wait to make his own contributions to theoretical physics. So whenever he had free time in the patent office, he took the opportunity to work on his own physical theories. From the notes he developed, he proposed a series of concepts vital to the history of modern physics.

The year 1905 was a splendid one for the young German scientist's career. Four of Einstein's technical articles were published, representing an extraordinary contribution to the field of modern physics. These included a paper on Brownian motion, using a statistical theory of atoms to account for the herky-jerky movements of small particles in a fluid, as well as an article on the photoelectric effect, an exposition of elementary quantum ideas that would win him the 1921 Nobel prize.

The other two articles published by Einstein concerned the theory of special relativity. Basically, he examined the effects of traveling close to the speed of light. Using the supposition that the speed of light is constant for all observers, Einstein calculated that the lifetime of a moving object, according to the vantage point of a fixed framework, would appear prolonged. He also found that the length of such a body would seem contracted in the direction of motion. For these revolutionary ideas, which were subsequently verified through experimental testing, Einstein won his greatest acclaim.

In order to explain special relativity in a simple fashion, Einstein developed the notion of expressing space and time as a single four-dimensional entity called space–time (three spatial dimensions plus one temporal dimension). Viewing time as the fourth dimension was not an entirely new idea—even H. G. Wells explored this concept in his science fiction—but Einstein was the first to codify four-dimensional space–time in a physically complete and mathematically rigorous manner.

For Einstein, time was not just another ordinary dimension, measurable by a meterstick using real (conventional) numbers. For him, time was imaginary. Not imaginary in the sense of fictitious but, rather, in the mathematical sense: expressible in terms of multiples of the square root of negative one. In other words, unlike the space coordinates, which squared yield positive quantities, the squares of the time coordinates are negative.

The result of such a system is that one can map out space–time "events" on a four-dimensional lattice and then define space–time "distances" between them. In this context, an event is something that happens at a particular place and time; for example, Mary drops a tray of dishes in the Main Street Diner, on the corner of Main and High Streets, at 5:35 P.M., and Xyzzelak waves a tentacle in his Mars-orbiting spaceship, 8000 kilometers above the planet, at 8:48 A.M. These are two different events, separated by both time and space.

Distances in space–time are defined by taking the square root of the sum of the squares of the spatial and temporal coordinates. This is just the Pythagorean theorem applied to the fourth dimension. But since the square of the time coordinate is negative, space–time distances possess a different character than purely spatial distances. While the latter can be only positive, the former can be either positive, negative, or zero. So, it may be the case that the space–time distance between Mary and Xyzzelak is negative. Finally, to complete this analysis, we define a mathematical object, called the *metric*, to represent the set of all possible space–time intervals between events. Plug in any two events and the metric yields a value for the interval between them. Now, this value could be either positive, negative, or zero, representing the causal relationship between the events concerned. In particular, according to convention, a value of zero for the metric distance indicates that the interval between two events is such that light could travel between them. In free space, light travels in a straight line; therefore, in special relativity, metric distances of zero are associated with straight lines.

From Mass to Motion

Imagine describing the history of the cosmos with just one equation. Not the nitty-gritty details, but the overall large-scale behavior of objects in space, as well as the ultimate fate of the universe itself. That is what Einstein's theory of *general relativity*, the successor to special relativity, essentially does.

Einstein's later theory, published in 1916, significantly augments his earlier concept; while special relativity deals with objects moving only at constant speeds, general relativity includes accelerating bodies as well. In particular, general relativity is a theory of gravitation and its effect on accelerating masses. In contrast to Newton's theory of gravitation, which depicts it as an attractive force between pairs of objects, general relativity represents gravity as a universal property of space. This representation can be achieved because gravity is a force that accelerates objects at rates independent of their masses. Hence, general relativity, as a global theory of both material and vacuum regions of the cosmos, makes full use of the notion that gravitational acceleration can be defined in the absence of mass. The theoretical bedrock of general relativity is a hypothesis known as the *Principle of the Equivalence of Gravitation and Inertia*, which indicates how any physical system behaves in a gravitational field. Briefly, it states that it is impossible for any measurement to be performed that distinguishes between an object freely falling under gravity and the same object residing in deep space without accelerating.

This principle is a natural extension of the well-known phenomenon that a freely falling object appears weightless. Imagine, for example, that a man named Icarus is trying to weigh himself inside an elevator. Unfortunately, as soon as he gets on the scale, the main cable breaks and the elevator plummets downward. Just before Icarus hits the ground, he glances at the reading and finds that he weighs zero pounds. His weight at that instant is the same as it would have been if he were unaccelerated in space, rather than gravitationally accelerated (being drawn toward Earth) in free-fall.

The Principle of Equivalence takes this one step further. All forces and bodies, all physical systems whatsoever, must display no difference between nonacceleration and free-fall. Thus, even the path of a beam of light must appear the same for an unaccelerated frame of reference and a frame that is falling freely under gravity. In other words, an astronaut freely falling toward Earth and an astronaut "parked" in deep space (experiencing no gravity) would observe a light ray passing by to follow the same course. This proves to be a most interesting result—as depicted in a famous Einsteinian thought experiment.

Mapping Our Fate

Imagine two identical elevators: one parked in deep space, the other freely falling in a shaft. Following the Principle of Equivalence, light shone through one elevator should take exactly the same route, according to an observer on board, as light shone though the other. Now suppose that a beam of light emitted from a horizontally aimed laser positioned on the left wall of the deep space elevator travels straight across to the same height on the right wall. Then, as viewed by someone inside, if an equivalent laser emits a beam of light from the side wall of the free fall elevator, the beam must also travel straight across.

So far this Principle of Equivalence argument seems entirely reasonable. From the point of view of observers inside the two elevators, it seems natural that both of the light beams should follow equivalent straight-line paths. However, if we stand outside the elevators and look at the events taking place within, a different picture of the light paths can be observed. From the external point of view there is a substantial difference between the deep space and free-fall behaviors.

If we were to position ourselves in deep space right outside of the first elevator, we would still see the laser beam travel straight across from left to right. But if we stood outside the free fall elevator, we would see the light beam curve as the elevator dropped. The curving would follow the motion of the elevator as it plummeted toward Earth (because from the elevator's point of view, the beam still travels straight across).

How does this bending arise? It couldn't be due to an effect of the free fall elevator itself, because elevators don't bend light. It must therefore be an effect of the relative positions of the two beams. While the first beam is in deep space, far from gravity, the second beam is experiencing a gravitational field. Thus, inevitably, the Principle of Equivalence leads one to the conclusion that gravity bends light.

It is a well-known fact that light, unimpeded, always follows the shortest possible trajectory: a straight line. If gravity bends the path of light, then there are two possibilities. Either light isn't traveling on the shortest route, or the shortest route is really a curve. Einstein, seeing no evidence for the former, chose the latter conclusion. He postulated that gravity must force the straightest lines of space into curves.

The most direct routes in space, the paths that light rays must follow, are called *geodesics*. The metric distances along these paths are defined to be zero. In the absence of gravity, the natural form for geodesics would be straight lines. But once gravity is added, geodesics in the vicinity of the gravitational field become curves. This requires an alteration in the metric; it must be transformed to account for a new set of distance relationships between points.

The mechanism to change the metric when appropriate, and hence transform straight lines into curves, is called Einstein's equation. Einstein's equation provides a way of depicting gravity by describing the way space is affected by mass and the way the paths of celestial bodies are altered by curved space. While the right-hand side of the equation depicts the distribution of mass throughout space, the left-hand side of the equation, a function of the metric, represents the behavior of space itself. Essentially, according to this relationship, matter bends space, and the curvature of space, in turn, dictates the movement of matter.

Metaphorically, gravitation can be described in terms of the stretching of a fishnet. Imagine a fishnet spread straight out over the surface of the ocean such that it lies flat on top of the water. This flat fishnet represents a region of

Mapping Our Fate

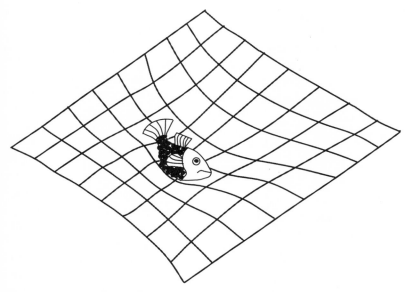

Figure 14. A fish caught in a fishnet. The distortion of the net due to the weight of the fish represents the influence of a massive body (such as the sun) upon neighboring geodesics. This deformation, in turn, bends the paths of objects traveling through this region, producing the effects that we identify as gravitational phenomena.

deep space, in the complete absence of mass or energy. The metric of the fishnet, describing the spacing between the strands, confirms that the net is flat. It also indicates that the strands are totally straight and evenly spaced, so light would follow a straight line if it traveled along any of them.

Now let us add mass to the fishnet. A jumping catfish leaps into the net, bending it down into the sea. Now its strands are no longer straight; they are curved very slightly. Next, a great white shark jumps in. The net sinks much farther downward, curving the strings considerably. Light, taking the shortest path along a strand, would no longer follow a straight line. Finally, a whale enters the net and it breaks. The whale simply rips through the center. This symbolizes the ultimate curvature of space–time—the escape-proof domain of a black hole. Black holes are so dense, and curve space–time so much, that not even light escapes their grasp.

The Power of Omega

As soon as Einstein developed general relativity, he and other researchers set out to find its applications to cosmology. They wished to know what relativity indicated about the origin, development, and fate of the universe. By utilizing the estimated density and distribution of the material and energy content of the cosmos, they sought the evolution of its spatial distance relationships as a result.

The first step in using Einstein's equation to determine the behavior of the cosmos is to break up space–time into space and time components. Although Einstein's theory represents space and time as two different parts of a whole, it is obvious that our experience of these parameters is quite different. Time flows ever forward, but space doesn't; therefore we wish to represent the space of the universe evolving over time. This is done by "slicing" up space–time, in a manner similar to carving a turkey, so that it becomes a temporally ordered collection of spacelike (spatially oriented) slices.

Another way of looking at this is to imagine that space–time is a roll of film containing the entire history of the universe. In order to understand how space evolves we can't simply look at the whole roll at once. To capture the essence

Mapping Our Fate

of a changing cosmos we must view the film, instant by instant, one frame at a time. This ensures that the laws of cause and effect, with which we are intuitively familiar, are conveyed; while a universe that happens "all at once" excludes the commonplace notion of causality, a universe that evolves over time preserves it.

Once the spacelike slices are cut, we must decide on the general conditions for their shape. This might seem a bit like cheating; isn't Einstein's equation supposed to determine the shape of the cosmos from its mass distribution? Unfortunately, Einstein's equation cannot yet be solved in its most general form. Solutions can only be found when the range of possible geometries is narrowed down a bit.

The most common supposition that those trying to unravel Einstein's equation make about the geometry of spacelike sections is that each slice is both homogeneous and isotropic (appearing the same in all directions). By applying this simplifying assumption—the Cosmological Principle—one slims down considerably the volume of conceivable universe models. Unless there is reason to look specifically at unevenly distributed cosmologies, almost all researchers utilize the Cosmological Principle in formulating their approaches.

The first homogeneous and isotropic cosmological solution was found by Einstein immediately after he developed his formalism. Einstein wasn't pleased by the result, to say the least. His model was unstable; it expanded until it reached a maximum radius and then contracted down to a point. The matter in his cosmology drove the spacelike sections to either grow or shrink; there was no in-between.

Hubble had yet to discover the universal recession of galaxies. The Big Bang theory had not yet been developed. No one knew that the universe was expanding and Einstein was mortified because his equations implied that it was. Therefore, he never published these initial results.

Why was Einstein so afraid of an unstable cosmological model? Having overturned large portions of Newton's theory, perhaps he was trying to preserve what he could of the Newtonian notion of fixed space. Or maybe it was his religious conviction that the force of nature is eternal and unchanging? At any rate, Einstein vehemently argued (at first) that the universe, as a whole, must be stationary.

In order to "correct" his cosmology, Einstein decided to amend his original equation in the least upsetting way he knew how. He added a term, called the cosmological constant, which served to counteract gravity (on the largest scale) and stabilize his model. Einstein announced his modified approach in 1917. Later, after learning of the Hubble expansion, he was to call his addition of the cosmological term the greatest blunder of his life.

In the period that followed, additional solutions were found to Einstein's equation, both with and without the cosmological constant. The Dutch astronomer Willem de Sitter, who during wartime helped to publicize Einstein's writings outside of Germany, found, in 1917, the second known solution. His model included the unnatural assumption that the universe was a complete vacuum. This supposition of emptiness led to a behavior of exponential expansion. Later, the inflationary universe theorists would modify de Sitter's model to approximate the idea of an epoch of rapid expansion.

Then, in 1922 the Russian mathematician Alexander Alexandrovitch Friedmann developed several dynamic cosmological models based upon Einstein's original equation without the cosmological term. Friedmann, who was born in St. Petersburg, spent a considerable portion of his career

Mapping Our Fate

as a meteorologist before becoming interested in the mechanics of relativity. Once he took to cosmology, however, he ended up outdoing Einstein in the quest for an accurate portrait of the behavior of the universe. Part of this was timing. A decade after Friedmann produced his work, Hubble discovered the universal redshift of galaxies and the expansion of the universe. Einstein's cosmological constant term became seen as unnecessary, and models, namely Friedmann's, which eschewed this term—and hence predicted universal expansion—were revived. As it turned out, Friedmann's models without the constant, rather than Einstein's with the constant, became the basis of the Big Bang theory and modern cosmology. Although modern inflationary theory ponders an exponentially expanding (de Sitter) phase of cosmic evolution, it is assumed that this era would be extremely brief and that the universe would rapidly switch over to a Friedmann sort of dynamics.

Specifically, Friedmann found three possible types of behavior for solutions of Einstein's equation, subject to the Cosmological Principle. These models, characterized by their spatial geometry, are called open, flat, and closed. As derived by Friedmann, the evolution of the universe is completely determined by the geometric category in which it falls.

An open universe possesses spacelike slices that are shaped like saddles (technically called *hyperboloids*), not three-dimensional saddles, but saddles curved into a higher dimension. As we'll discuss, the curvature of space is not a property that can be viewed in any of the conventional dimensions, but instead could only be seen by an observer living in a higher-dimensional realm. Since this domain is obviously inaccessible to us, we can only imagine what such curvature might look like. Using a three-dimensional ana-

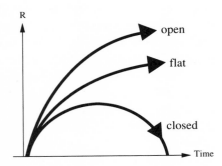

Figure 15. The evolution of the universe for the three basic Friedmann models. Here "R" represents the overall scale of the cosmos, and the horizontal axis represents its age. Note that in just one case, the closed model, does the universe recollapse in a finite amount of time.

logue, we imagine a surface that is curved downward in two directions (the sides of the saddle) and upward in two (the front and back of the saddle). Because these sides and ends never meet, the scale of open models is infinite. One could never circumnavigate a cosmos of this sort.

As classified by their behavior, open universes have the property that they begin expanding from a point (or from an inflated bubble, in the case of inflationary theories) and continue to expand forever. In all Friedmann models, a single scale factor, called R, characterizes the relative dimensions of the universe for any given time. In the case of an open cosmology, R is a continuously growing entity; it never gets smaller.

External expansion, though not exactly in the same manner, is a property possessed by flat models as well. R tends to grow in a flat universe, but not as rapidly as in an

Mapping Our Fate

open cosmology. As a flat cosmos ages, its growth tends to slow down, but never quite stops or reverses.

Flat universes are said to possess Euclidean spatial geometries (the standard, noncurved geometry learned in school). In other words, the spacelike slices of these cosmologies are the equivalent of planes, extended to a higher nonvisible dimension. Just as in the case of open models, if you traveled along the infinitely long hypersurface (higher-dimensional plane) of a flat geometry, you'd never reach your starting point.

Closed universes, on the other hand, are entirely circumnavigable. They possess hyperspherical (four-dimensional sphere) geometries, enabling one to reach one's starting point by continuing indefinitely in a straight line. Thus the extent of space in a closed universe is manifestly finite.

Dynamically, closed models present us with the modern scientific equivalent of the Hindu Kalpa or the Babylonian Great Year. Space starts out as a point (or bubble), then expands outward as a continuously growing hypersphere. But then, in contrast to both open and flat universes, the expansion slows down, stops, and finally turns into contraction. As the scale factor R reverses its behavior, the universe proceeds to undergo a complete collapse, right back down to a point. What happens then is more speculative. Many believe that another Big Bang would be reenacted and that the expansion and contraction phases would take place again and again, in cyclical procession.

It is unclear to present-day physicists which of these models represents our own universe. But fortunately there is a clear way of deciding this question. If the average mass and energy density of the cosmos is *less* than a certain value, called the critical density and estimated to be about 10^{-29}

grams per cubic centimeter (the exact number depends on the current value of a quantity known as the Hubble parameter), then we live in an open model that will expand forever. If, however, this average is *equal* to its critical value, then we reside in a flat space. And finally, if the average density is *greater* than critical, then we can be sure that the cosmos will someday collapse.

These relationships can be quantified by use of a quantity called omega: the ratio between the actual and critical densities. The appropriate Friedmann model is completely determined by omega. If omega is less than one then the open model is appropriate; if omega is equal to one, then the flat model must be used. And, if omega is greater than one, then we live in a closed Friedmann cosmology.

This distinction can be seen by use of a rocketship analogy. Imagine that a rocket were launched from Cape Canaveral with only the bare minimum of fuel, just enough to leave the ground. The ship would take off, but would soon reverse course and return to Earth. This scenario is similar to the closed Friedmann case, where the universe cannot reach its own escape velocity, and must eventually recontract.

Now, in contrast, imagine that the rocket left full of highly powerful fuel. It would blast off and continue to move upward at rapid speed. Soon it would find itself far beyond the pull of Earth's gravitational field. If unimpeded by other celestial objects, it would travel through space forever. This is analogous to the open universe model, where the cosmos exceeds its own escape velocity, and hence continues to grow forever.

Finally, picture the in-between case. A spaceship has enough fuel to blast through the atmosphere, but not quite enough to leave the Earth's domain. It settles, then, into a comfortable orbit around the Earth. This is the terrestrial equivalent of Friedmann's flat universe.

Might there be other types of universes that obey the Cosmological Principle and satisfy Einstein's equation without the cosmological constant term? Apparently not. In 1936, H. P. Robertson and A. G. Walker proved that the metric used in the Friedmann models constitutes the most general for an isotropic and homogeneous space with no cosmological constant. No other models, with these standard specifications, can be constructed. (Belgian cosmologist George Lemaître used Einstein's discarded cosmological constant, however, to create an alternative, but little used, set of models.) Therefore, these prototypical cosmologies, classified by the terms open, flat, or closed, are usually called Friedmann–Robertson–Walker (FRW) universes. They constitute the mathematical framework for the Big Bang cosmological approach.

The Price of Inflation

How can we determine omega and decide if our universe is open, closed, or flat? A host of methods are being employed in trying to resolve this issue—to determine if the universe will one day undergo recollapse. A group of theorists, the inflationists, think they already know the answer. One of the most confident predictions of the inflationary universe model is that omega is equal to one. Thus, those believing in inflation as currently formulated think that the universe is flat.

This prognostication stems from an estimation of the remarkable flattening properties of the inflationary phase, as built into the model. The theory of inflation is specifically designed, in fact, to smooth out any initial curvature of the universe. The reason for this is that one of the motivating factors for the creation of the inflationary model was to solve the so-called flatness problem.

The flatness problem was pointed out by Robert Dicke and fellow Princeton professor Phillip James Peebles in the late 1970s. It derives from early estimates of the modern universal value of omega. At the time of Dicke and Peebles, experimental measurements indicated that the contemporary value of omega fell within a range of .01 and 10. This lack of certainty reflected a dearth of physical techniques for measuring omega. Today we can do much better and narrow this range of values.

The question posed by Dicke and Peebles pertains to the value of omega in the very early universe. In the standard Big Bang model, it can be shown that omega tends to grow (if it is above one) or shrink (if it is below one) as time passes. Therefore, for the value of omega to fall within the accepted range today and be so close to one, it must have been incredibly close to one in the early universe. For example, at a time of one second after the Big Bang, omega must have been equal to one to an accuracy of one part in a quadrillion (thousand trillion).

The inflationary universe theory purports to eliminate this problem by assuming that all initial bumps and curves are stretched out during the de Sitter phase of exponential expansion. Therefore, if one assumes that an inflationary era took place before the Friedmann phase began, one might imagine any possible value for the curvature of the very early universe; it all becomes smoothed out in the end.

Picture an early era of the universe in which the overall energy density is much less than the critical density. There is a strong negative (saddle-type) curvature associated with this discrepancy. Now imagine that an epoch of extraordi-

Mapping Our Fate

narily rapid growth takes place. The cosmos becomes inflated so much that any bumpiness becomes stretched out considerably. Soon any sort of noticeable curvature is flung out of the region that is destined to be the currently observable universe; only flat space remains within these limits. By the time the ordinary Hubble expansion begins, the cosmos has already become a flat Friedmann model.

Inflationary theory was devised, in part, to explain why omega falls within .1 and 10. It doesn't *assume* that the universe today is flat. But, according to this model's mathematical consequences, omega today must necessarily be equal to one exactly (or at least extremely close to one). This is the strongest prediction of inflation and is currently being tested. If omega is indeed precisely one, this will be an enormous coup for inflation. But if the universe, instead, is closed or open, rather than flat, then inflation will need to be significantly revised. Experimentation will likely yield a bounty of results to resolve this issue once and for all.

Inflation, with all of its obvious virtues, is by no means the only solution to the flatness and horizon problems. Other attempts to resolve these issues have come and gone—some falling flat when experimental verification has been tried; others not yet tested by the rigorous constraints placed by astrophysical data. Many researchers have suggested that the cosmological paradoxes might be resolved by simply assuming, as a given, that the universe started out as perfectly isotropic and homogeneous. In that case, the regularity of the cosmos would be "built in" as an initial condition, rather than as a result of an inflationary epoch. Naturally, if omega turns out to be significantly different from one, there will be a host of alternative theories popping up to provide a suitable explanation of the dynamics of the very early universe.

5

The Big Crunch

The fate of the cosmos has been a subject of speculation for thousands of years, but only now do we have the mathematical tools to resolve this issue, at least potentially. Einstein's equation and its most elegant set of cosmological solutions, the Friedmann models, have proven most effective in suggesting a well-defined evolutionary chronicle of the universe. This has led to the standard model of universal expansion, the Big Bang theory, which has been confirmed by numerous experiments, including Hubble's telescopic observations of galactic recession, Penzias and Wilson's detection of the cosmic background noise, and most recently the Smoot team's detailed measurements, using COBE, of cosmic microwave background radiation. Time and time again it has been demonstrated that the standard model is a highly accurate means of describing the content, character, and developmental history of the cosmos.

The main factor in completing this picture is a precise assessment of the value of omega, a determination that will provide crucial information about the mass, the curvature, and, hence, the destiny of the cosmos. Will the universe develop forever, that is, until its state of maximum entropy is reached? In that case the end of the cosmos would be drawn out. Slowly, but surely, all of the stars in the sky would turn into black holes, neutron stars and other forms of charred, lifeless stellar relics. Eventually, all the usable energy would be exhausted and universal heat death would ensue.

Or will the universe contract down to a point once more in a colossal Big Crunch? The case of closed space would portend a far more exciting and definitive demise of the universe than the open model would indicate; the end would come with a bang, not a whimper. In a closed universe, everything that we know—all matter and energy—would necessarily be reprocessed into a minuscule state once again in a sort of time-reversed Big Bang, at least in appearance. And then, perhaps, the universe would start anew in another cycle of time.

Omega is the missing link in divining which it will be: open or closed, infinite or finite, forever expanding or cyclical. Let us now look at the many ways in which experimentalists are currently trying to determine the value of this vital quantity.

THE SHAPE OF CREATION

I love cosmology; there's something uplifting about viewing the entire universe as a single object with a certain shape. What entity, short of God, could be nobler or worthier of man's attention than the cosmos itself? Forget about interest rates, forget about war and murder, let's talk about space.
-Rudy Rucker, The Fourth Dimension

Longevity Tests

As the number of octogenarians and nonagenarians in our society has continued to rise, due, in part, to better health care and nutrition, the interest has grown in the science of longevity. People want to know what they can do to prolong and enrich their lives. Increasingly, large segments of our culture are turning to fitness centers, health clinics, sports magazines, diet advice agencies, workout videos, and the like, anything to gain a few more years of earthly existence.

If a doctor is asked about what one might do to increase one's chances for a longer life, there are a number of responses he or she might give:

"Cut back on fatty desserts."

"Stop smoking."
"Take long walks at least three times a week."
"Don't drink to excess."
"Avoid thinking too much about cosmological paradoxes."

Naturally, there are no guarantees that any of this advice would work. After years of careful dieting and exercise, one might end up fatally injured by slipping on a banana peel dropped on the floor of a natural foods restaurant. But still, as any insurance broker or actuary would tell you, there are statistical ways of assessing one's lifestyle to predict one's lifespan.

Imagine if the universe were to walk into a health clinic and ask, "Doc, how long do I have to live?" The doctor (a clinical cosmologist), after getting over his shock at finding all of physical reality in his cramped office, would likely ponder for a moment and say, "Well it depends on a physical parameter that we call omega. If omega is small enough, you can live and grow forever. Otherwise, you'll eventually crunch up."

"Put it to me in plain English, Doc. I don't understand these technical terms," the universe might reply.

"Uh right. Let me give it a try. Your fate, my dear cosmos, is determined by your weight, so to speak. I don't know your mass, pal, but if you are too heavy, according to my charts, I predict that you will collapse some day. On the other hand, if you are slim enough, and remain below a certain critical mass value, you don't have to worry about total collapse. You might lose all of your available energy, but you'll always exist."

"Collapse? Lose my energy? Doc, what can I do to stay fit and expand my horizons forever?"

"Sorry, there's nothing you can do at this point. Your destiny was fixed at your moment of birth, when you

Shape of Creation

emerged from your womb of nothingness. But, to spare you any further anxiety, I can perform some tests to find out your omega value right now. Then, after appropriate counseling, you will eventually learn to accept your fate."

"What kinds of tests? Will they hurt?"

"Not really. There are several kinds of tests, all quite painless. One indication of your omega value would be the results of a *shape test*. If you are plump like a sphere, then you are bound to suffer collapse. However, if you are shaped like a plane or saddle, you don't have to worry about implosion. I'd have to take some curvature measurements on you to figure out into which category you fall.

"Then, there's the *expansion rate test*. I'd need to determine how big you are and how fast you are growing. This should tell me whether or not your expansion will eventually slow down and reverse itself. There's a second type of *motion test* that is less direct. You would have to spin around and I would assess your mass by the way that you move. From your mass, I would decide if your omega parameter is above or below its critical value, one.

"Finally, there's the *constituent test*, where I would figure out what you are made of. I'd need to know exactly what types of particles are inside you, and the mass of each constituent. This may be difficult if most of your matter is invisible. Based on these quantities, I could then estimate your total mass, which would, in turn, yield omega."

"Doc, let's try the shape test first. It sounds the easiest."

Getting in Shape

The idea of space having a particular shape is fairly recent. Before the 20th century, space was thought of as merely the absence of geometry: a formless void. Those astute enough at that time to imagine sending measuring gear into space would still have presumed that all distances and angles surveyed would follow precise Euclidean form in their relationships.

Euclid's geometry, developed more than two millennia ago in classical Greece, prescribes a number of well-defined rules relating various properties of points, lines, and figures. For example, any two points on a plane determine a single straight line. Also, for any given line and any given point outside of that line, there exists a single (second) line through the point parallel to the first line. It can be further shown that these two parallel lines never meet. This is called the parallel axiom.

The Euclidean concept of a triangle is another case of strictly constructed principles. The definition of a triangle is clear: three sides and three angles. Moreover, the sum of its angles is equal to 180 degrees; no more, no less. Furthermore, if one knows the lengths of any two sides and the number of degrees of the angle that links them, then the length and values of the remaining sides and angles can be exactly determined.

Although Euclidean geometry works well for objects on a plane, it falls short in describing the relationships of distances and angles for figures on a curved surface. Therefore, so-called non-Euclidean geometry, the geometry of curved space, was developed for nonplanar situations. This is the mathematics used by Einstein in his theory of general relativity.

The first outline for a non-Euclidean geometry was developed by the 19th-century mathematical genius, Karl Friedrich Gauss. Gauss reexamined Euclid's work, and tried to simplify it. To his surprise, he found that he could replace the parallel axiom—considered almost sacred by mathema-

Shape of Creation

Figure 16. Karl Friedrich Gauss, 1777–1855. (From Albert Betlex, *The Discovery of Nature*. New York: Simon and Schuster, 1965. Courtesy of AIP Niels Bohr Library.)

ticians—with an alternative set of principles relating points and lines. For instance, he showed that, under certain circumstances, parallel lines could meet. When all was said and done, he had created an entirely new system, a complete non-Euclidean geometry. Not wishing to appear too unorthodox, he never published his work in this area. It was found after his death, buried in his papers.

In the 1830s, before Gauss's non-Euclidean theories became known, two other European mathematicians, Nikolai Lobachevsky of Russia and János Bolyai of Hungary, each published his own non-Euclidean model of space. Unfortunately these important bodies of work were largely neglected for about 30 years until Gauss's writings on the subject were finally published posthumously.

In the models of Bolyai and Lobachevsky, geometric figures, such as triangles, squares, and other polygons, possess unusual properties. For instance, the sum of the angles of a triangle is always less than 180 degrees. Also, the larger the area of a triangle, the smaller the sum of its angles. Conversely, the smaller the area, the closer the angular sum to 180 degrees. So for very small triangles, it is difficult to tell the difference between the Euclidean and non-Euclidean pictures.

Soon after Bolyai and Lobachevsky's work appeared, another non-Euclidean approach, with distinct predictions, was developed by the German Bernhard Riemann. Riemann, who dropped out of the ministry to study math, much to the chagrin of his father, a Lutheran pastor, found a new set of axioms and theorems. In particular, he found that Euclid's parallel axiom could be replaced with a postulate that would allow parallel lines to converge at a point. In other words, instead of two points determining one and only one line, they define an infinity of intersecting lines.

Riemannian triangles stand in contrast to the Bolyai–Lobachevsky sort. They possess angular sums in excess of 180 degrees; three right angles are even possible. Moreover, the greater the size of a triangle, the larger the discrepancy with the Euclidean form. That is, bigger triangles have greater angular sums.

Where might we find parallel lines and triangles of the sorts described by Bolyai, Lobachevsky, and Riemann? The answer lies in an analysis of curved space. Space of negative (hyperbolic) curvature possesses the non-Euclidean geome-

Shape of Creation 163

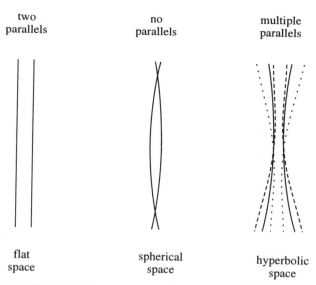

Figure 17. Parallel lines in various geometries. In Euclidean geometry (flat space), each point outside of a line defines a single line parallel to the first. In Riemann's model (spherical space), "parallel" lines converge at a point; there are no real parallels. Finally, in the geometry of Bolyai and Lobachevsky (hyperbolic space), each point outside of a line defines an infinite number of parallels to that line.

try of Bolyai and Lobachevsky. Space of positive (spherical) curvature can be best described by Riemannian geometry. Thus one can, at least in theory, readily determine the curvature of a region by examining the geometric properties of lines and figures included in that region.

Einstein's theory of general relativity borrowed heavily from the work of the non-Euclidean geometrists. In order

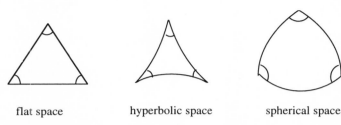

Figure 18. Triangles in various geometries. Whereas in flat space the sum of a triangle's angles must equal 180 degrees, in hyperbolic space the sum must be less than 180 degrees, and in spherical space the sum must be greater than 180 degrees.

to model gravity, he utilized a higher-dimensional version of non-Euclidean geometry. Essentially, he found that five dimensions were needed to describe the gravitational properties of space-time: three to model space, one to model time, and an additional nonphysical dimension to accommodate curvature.

This fifth dimension is quite peculiar. The laws of physics operate perfectly well in a four-dimensional space-time; there is no need to postulate a fifth dimension to explain any observable properties of nature. Yet because space-time is curved, it must be curved into something, and that something is the fifth dimension. We do not know if this higher dimension has any physical significance, only that it is a mathematical necessity.

How can we understand the concept of higher dimensions if we cannot perceive them? If we stand in a corner, we can picture the first three dimensions as the two walls and floor that come together at our feet. If we move out of this position, then we can imagine that the fourth dimension

Shape of Creation

represents the duration of our motion. These ideas are fairly straightforward.

However, in order to comprehend the curvature of this space-time continuum into something else, we cannot use our intuition. Instead, we must construct an analogy, a lower-dimensional model of higher-dimensional reality. Fortunately, such an analogy has already been constructed for us: the fictional world of Flatland, described by Edwin Abbott in a book of the same name, published in 1880.

The View from Flatland

Imagine a world in which there are only two traversable dimensions, North-South and East-West. Up and down do not exist at all as independent entities; in this world, for all intents and purposes, they are equivalent to north and south. One cannot look, hear, travel or send signals beyond this two-dimensional realm. Aside from the surface of Flatland, all of space is completely inaccessible.

The creatures of Flatland are geometric figures, such as line segments (women), polygons (men), and circles (priests). There is a strict caste system, with citizens segregated according to their numbers of sides and angles. Triangles have the misfortune of being the lowest male members of the hierarchy, and women, possessing no angles, are practically considered chattel. Successive generations accrue additional angles. This gives triangular fathers hope for the prestige of having square sons, pentagonal grandsons, and hexagonal great-grandsons.

The hero of Flatland is a square, who one day encounters a three-dimensional spherical creature. This being appears to him, at first, as a tiny point. Gradually, this point grows bigger and transforms itself into a growing succession of circles. From the creature's point of view, it has lowered itself into the plane of Flatland, exposing more and more of its girth. But the square cannot view this process in its totality. Because of his dimensional limitations, he can perceive an evolution only within his own plane. Only when the being lifts the square up out of the surface of Flatland does the truth become known to him. The square realizes, for the first time in his life, that the third dimension exists. Determined to spread the word, he encounters resistance wherever he goes. Ultimately, the square—now a social pariah—becomes resigned to the fact that only those who have experienced the third dimension can truly understand it.

Abbott's tale can be read as a parable of our own lack of knowledge about higher dimensions. Confined to a three-dimensional space, it is almost impossible for us to visualize a fourth directionality. Yet Einstein's theory requires us to do just that. Like the square in Flatland, we are forced to confront our own limitations in this domain.

Even more illustrative of the difficulties in picturing Einsteinian space-time is a book that appeared in 1965 as a sequel to *Flatland*. *Sphereland*, by Dionys Burger, picks up where *Flatland* left off, and extends Abbott's fable to new realms of social and scientific satire. It chronicles the life of a hexagon, the grandson of the now-venerated square, in his attempts to fathom his two-dimensional cosmic arena.

Society, in the time that *Sphereland* depicts, has come to see the error of its ways and has accepted the teachings of the square about a third, perpendicular direction. Along with this scientific paradigm shift has come a massive social revolution. Women are now basically respected, and the caste system is breaking down.

Shape of Creation

One set of dogma that has not been confronted by the Flatlanders, however, falls within the realm of geometry. Euclidean geometry, with its detailed axioms and theorems regarding triangles and other figures, remains entirely unquestioned. The hexagon, in collaboration with a mathematician named Mr. Puncto, decides to put these theorems to the test by measuring and summing the angles of triangles, comparing each result to the expected value of 180 degrees.

When they look at the angular sums of small triangles, they notice little discrepancy. But when they examine larger and larger figures, they are bewildered by greater and greater totals. The only conclusion that they can draw is that their space is not flat at all; rather, it is spherical. The positive curvature of their world causes its geometry to be measurably non-Euclidean in its character. The Flatlanders realize that if one were to set off in a particular direction, one would eventually reach one's starting point.

At the end of *Sphereland*, the characters make a second monumental discovery. By setting out into "space," the two-dimensional vacuum between various Flatland "planets" and "stars," they find that they live in an expanding universe. In a scenario similar to Hubble's findings, they determine that all of the other worlds are uniformly receding from theirs. They fail to find out, though, whether Flatland once had a Big Bang or whether it may recollapse in a Big Crunch. Apparently, these difficult cosmological issues have been left to a potential second sequel.

The Fifth Dimension

When the people of Flatland discover, through angular measurements of large triangles, that their world is actually spherical, there is a considerable amount of confusion. Unlike a plane, a sphere has a finite surface area. Normally, one assumes that a finite region possesses both a boundary and a center. Yet, within its own two-dimensional realm, a spherical surface has no boundary nor center (its center lies *within* the sphere, not on its surface).

When we consider the Friedmann cosmological models, we are presented with a similar quandary. Two of the Friedmann cosmologies, the open model and the flat model, are infinite in extent. The third Friedman cosmology, the closed model, possesses positive curvature, however, and is therefore considered finite. The question then arises for this third sort of space-time, if it is finite, then where are the boundaries and center?

For many, a finite universe is more philosophically appealing than a infinite model. It is easier to picture a finite number of galaxies than an infinite number. And it is certainly more gratifying to posit that Earth's domain comprises an extremely small, but nevertheless finite, portion of all existence rather than an infinitesimal, and ultimately insignificant, pointlike speck.

However, it is a common misconception, for those considering a finite universe, to imagine that there exists a boundary somewhere in space, beyond which no one might cross. I've been asked many times: If space is finite, then what would happen if you crossed over its walls? Would you see more space or would you see nothing at all? Or would guard dogs and mine fields block you from venturing into this "forbidden turf?"

A related question that is also asked of me frequently by those pondering a finite, spherical universe is: Where is the center of space? I suppose that many of those making this query imagine that there is a sort of "North Pole" somewhere in the cosmos, from which all matter has emanated. If we could travel to this pole, according to this logic, we would be equidistant from all parts of the periphery of space.

Such a wall or pole cannot exist, however. The existence of either a spatial boundary or spatial center would directly contradict the Cosmological Principle by assigning the Earth a special place in the cosmos. For if such special regions did exist, then it would be theoretically possible to determine the relative centrality of Earth, that is, how close we are to the center and how far we are from the boundary. Consequently, if this were the case, the sky would no longer be isotropic and homogeneous. Since all indications from COBE and other sources do validate the cosmic isotropy and homogeneity associated with the Cosmological Principle, then we must necessarily preclude the existence, in physical space, of a central point or universal boundary.

Assuming that the universe is a hypersphere (higher-dimensional sphere), it can be said to have no boundaries at all. Traveling in any chosen direction, we would eventually reach our starting point. Like the characters in *Sphereland*, we must also suppose that the center of the universal hypersphere lies outside of its spatial domain, situated in a higher dimension perpendicular to our own. Therefore, only if we were able to access the fifth dimension, might we ever hope to visit the true midpoint of the universe.

Beyond the Fifth Dimension

If the universe has five dimensions, including space, time, and the nonphysical direction of curvature, then why not more?

Shortly after Einstein published his gravitational theory, the German mathematician Theodor Kaluza asked this very question. In his search for a unified field theory of gravitation and electromagnetism, he proposed a model in which general relativity is extended by one more dimension. This extension is performed to accommodate the components of the electromagnetic field.

To account for the fact that such an extra dimension cannot be directly observed, Kaluza imposed the additional constraint that it be small, circular, and hence undetectable. In order for it to be unseen, its circumference would have to be comparable in size to the Planck length, approximately 10^{-33} centimeters.

Why wouldn't a tiny, cyclical dimension be noticed? A circular dimension would be like a conveyor belt, whisking every point in space around a closed loop. Imagine stepping on such a belt. If it were large you would notice traveling around a circle along a higher dimension. But if it were extremely small, you wouldn't even notice the difference; you'd be essentially at the same place all the time.

The justification for this strange condition of microscopic circularity came almost a decade later, in 1926, when Oskar Klein noted that the newly emerging theory of quantum mechanics contained numerous examples of periodic boundary conditions. Klein therefore proposed that the higher dimension was "quantized," meaning that it was limited to a minute cyclical range. No particle could detect this circular path, because its circumference lay below the smallest observable unit (roughly the Planck scale).

Inspired by Kaluza and Klein, Einstein became fascinated by the idea of unifying the forces of nature. He wished to unite not just gravity and electromagnetism, but the nuclear forces as well, into a single higher-dimensional model. Much to his disappointment, as hard as he tried, he never achieved a successful model of unification. Nevertheless, though many theoretical difficulties need to be ironed out, "Kaluza–Klein theories" of the unification of nature's forces, within the context of a multidimensional space, remain quite popular.

In many of these Kaluza–Klein models, a process called *compactification* is proposed to have occurred in the very early universe (even before inflation). Originally, according to these approaches, all dimensions were of equal scale. The three dimensions of space and the higher dimensions associated with electromagnetism and the other nongravitational forces were all about the same size. However, at some point in the distant past, all of these extra dimensions shrank considerably, as the three spatial dimensions began to grow. Eventually, as compactification was completed, all higher dimensions were fixed at scales comparable to the Planck length. By contrast, the three spatial dimensions, would continue to grow to enormous scales, in the sort of Hubble expansion that we observe today. Thus, in summary, according to these Kaluza–Klein compactification theories, the expansion of the physical universe is symptomatic of the primordial collapse of unseen higher dimensions.

We have taken a bit of a leap, outside of the cozy confines our own familiar space–time, and into the lofty realm of multidimensional physics. Let's return from our higher-dimensional excursion—unbruised, I hope—to reconsider the question of our ultimate fate. In particular, let's examine how we might use our knowledge of space's geometry to determine its ultimate evolution.

The Cosmic Deficit

Before the invention of airplanes, balloons, and spaceships, the human race lived in a sort of Flatland. Not that the globe was any less three-dimensional than it is today. It was very difficult, though, at that time, to step outside of Earth's surface and view its overall geometric structure. It took the circumnavigation of the world to prove that it was round.

However, as the book *Sphereland* has shown us, even if flight were never invented and Columbus never sailed, we would still have been able to measure the Earth's curvature. By constructing a large enough triangle, we would have measured an angular sum of greater than 180 degrees, and thus concluded that the Earth's surface was positively curved.

Consider, for example, a great triangle formed by three terrestrial geodesics (straight lines). Take one of these lines to be the equator, another to be the Greenwich meridian (the line of longitude that passes through London), and the third to be Chicago's line of longitude. These three segments form a triangle with approximately three right angles (we could readily measure these by standing at the North Pole and at the two intersection points along the equator). Therefore, because 90 degrees plus 90 degrees plus 90 degrees is much greater than 180, we know for a fact that the world is spherical.

Now suppose we wished to apply this technique to the cosmos as a whole. We would first record the angle between two distant galaxies. Along with a third segment connecting these two galaxies, a triangle would be formed with our own. The angles of this triangle would change over time, so we'd need to record the exact moment of our measurement.

Next, we'd need to contact inhabitants of these two galaxies to find out the remaining two angles. The members of each galaxy would be asked to report to us the angle, from their perspective, between the Milky Way and the other galaxy. These measurements would all need to be performed

Shape of Creation

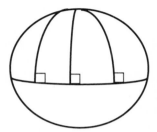

Figure 19. Lines of longitude intersecting earth's equator. Note that several triangles are formed, with the equator and lines of longitude, that have angular sums greater than 180 degrees. This is a feature of the spherical geometry of Earth's surface.

at the same instant, so a considerable amount of coordination would be needed.

Then, all of this angular information would have to be transmitted back to Earth. Because of the enormous distances involved, this would take millions of years, so we'd need to be extremely patient. Finally, in the far future, we could sum up the three angles and compare to 180 degrees. Depending on whether the sum was less than, greater than, or equal to 180 degrees, we would know whether the universe was open, closed, or flat. Obviously, unless we discover the secret of how to survive for thousands of centuries, this would not be a very practical method of measuring the cosmic curvature.

Another direct means of determining the curvature of the universe, and hence ascertaining whether or not omega is greater than one, is called the *standard candle test*. This method relies on the fact that galaxies in a flat space would appear, on average, to emit the same number of light rays within a particular angular spread; curved space, in contrast, would alter this amount, increasing it for a closed cosmos and decreasing it for an open cosmos. Therefore, the brightness of a galaxy can be compared to that of a standard candle (the typical brightness for a given distance), and an assessment of curvature can be made.

Let's examine how this technique would work in practice. Consider the light rays given off by a typical galaxy. In flat space, these beams of energy would dissipate into space, in all directions at equal rates, unhindered by the effects of gravity. Therefore, because the galaxy is assumed to be an average one, the amount of light emitted by the galaxy, as recorded by telescopes on Earth, would presumably be close to the normal value. Astronomers comparing the measured angular distribution of the galaxy's candlepower to the normal range would find it to be standard as well.

Now suppose that the universe is closed. Accordingly, all regions of the cosmos would have positive curvature, causing all spatial geodesics to be curved. As a consequence of this, the light emitted by the galaxy would be more focused, in comparison to the flat space scenario, and more rays would be detected per unit solid angle (a solid angle is the three-dimensional equivalent of a standard angle). Telescopes on Earth would therefore measure the brightness of the galaxy to be greater than the expected amount. Moreover, because the focused beams traced back would appear to have emanated from a larger source, the galaxy would appear broader. Its image would occupy a larger solid angle than normal. Thus, a galactic image that was brighter and more spread out than usual would be a telltale sign that space was hyperspherical.

If the cosmos is open, the effect would be the exact opposite. The negative curvature of space would serve to fan out all light beams from the galaxy, and there would be less measured radiation per solid angle. Also, if these fanned-out rays were traced back, they would appear to have come from a smaller apparent source. As a result, the image of the galaxy would seem dimmer and narrower than expected.

These techniques rely on a detailed knowledge of the absolute brightness (actual amount of radiation given off) of galaxies. Therefore, astronomers taking these measurements like to use galaxies with properties that are fairly well known. Typically, the giant elliptical galaxies that inhabit the centers of many clusters are utilized. They are prized for their almost uniform brilliance, and represent good standard candles for astronomers' purposes. The only drawback to the use of giant ellipticals is that their evolutionary properties are not yet well known. Because they probably lose brilliance as they age, we would need to know how old they are before we measure their light output. Also, they may gain in brightness by colliding with and absorbing other galaxies. As a result, the information obtained by this method is not yet that accurate. Early indications have pointed to a closed universe, and it remains to be seen what future observations will yield.

Direct measurement of curvature is an intriguing way of determining whether or not the universe will eventually recollapse, but unfortunately it isn't the most reliable method. Too much is unknown about deep space to expect that the angular distortions of light that we measure are necessarily produced by curvature. Because each celestial body is in constant motion and continuously evolves through aging and accumulation of mass, it is difficult to distinguish between the effects due to the curvature of space, and the influence of other dynamic elements, on light paths radiating out from an object.

In the next two chapters, we'll look at a number of more promising approaches, based on more reliable astrophysical methods. First we'll examine the ways researchers are trying to determine the deceleration parameter of the universe, the rate by which the expansion of the cosmos is slowing down. It is integrally connected with the more familiar parameter, omega, which measures the average density.

GALACTIC SPEEDING TICKETS

It is interesting to see that in the 50 years of observational cosmology following Hubble's first calibration of the extragalactic distance scale, our estimates of ages of things like the Earth, the solar system, the oldest stars and the universe itself agree reasonably well. If you go home and tell your children that the universe began 15 billion years ago, you won't be far from wrong. Hopefully the Space Telescope, now named after Edwin Hubble, will halve or quarter our errors.
-**John P. Huchra, On Contemporary Observational Cosmology: When Did It All Begin?**

The Robot Pitcher

If you, as an amateur, throw a baseball high into the air, it will rise up a few meters, stop, and eventually come down (not on your head, you would hope). A professional all-star baseball player might fare better, seemingly defying gravity with an exceptionally high pitch. But obviously no one on Earth can throw a ball high enough that it would rise above the clouds, let alone leave Earth's orbit.

Imagine, though, an android with the combined powers of Superman, Terminator, Bionic Woman, and Babe Ruth wielding a plutonium-driven titanium pitching arm with ultraefficient rocket engines. Grabbing a baseball, the android winds up, pitches, and it's out of here. Over the fence, beyond the clouds, in fact it has gone into outer space.

Compared to this robot, any earthly major league player would be no match; the human would get clobbered. Therefore, our bionic buddy is equipped with adjustable controls. On the lowest setting, he is the equivalent of a first-rate professional, easily beating any eager young strong-arm, but still keeping his pitches reasonably slow and within the strike zone. Set the controls higher, and the ball literally goes into orbit (this is useful for promotional displays). At the highest setting, it's bang, zoom, off into space.

The control box settings are labeled by a gauge marked "omega" and an arrow pointing to various figures. If the omega dial is set at less than one, then the robot's pitch is said to be "open," because any ball thrown would engage in an open path out of Earth's gravitational domain. If it is set at greater than one, then the "closed" pitch sends any ball back to Earth. A third, "flat" option, an omega dial setting equal to one, is used for throws that hurl baseballs into orbit. Cosmologists studying the actions of this robot control box immediately notice a close analogy with the three possible evolutionary behaviors of the universe.

One sunny day, the android pitcher is brought to the middle of Yankee Stadium in New York for a public demonstration of its skills. At the last minute, the robot's designers notice that something is severely amiss. The omega dial has broken off and the android is hurling baseballs at random speeds. Although all of these projectiles seem to be flung into the upper reaches of the sky, it is unclear which of them will land, which will orbit, and which will escape into space.

In order to allay concerns about errant landings, scientists in the stadium quickly devise a scheme to determine the ultimate fate of these baseballs. Their scheme is based on what they call the *deceleration parameter* of projectiles. All objects fired into the air must continuously slow down under the influence of Earth's gravity. If this deceleration (rate of slowing down) is large enough—the case for familiar projectile motion—an object will sooner or later fall back to Earth. However, rocket ships and other high-speed volleys, once they travel high enough, begin to decelerate at a slower rate, and end up either circling Earth or soaring through outer space. Therefore, by measuring this deceleration rate one can readily assess the "omega value" of a baseball, and predict whether its path will be open, flat, or closed.

To determine the value of the deceleration parameter, scientists use high-powered binoculars to take careful readings of the upward velocity of a baseball for each height that it rises. By plotting this information on a graph, they soon find out, for any given ball, if it is slowing down enough so that it will return to the ground again. In short, the robot's omega reading for any given pitch can be computed from the deceleration parameter of the baseball thrown, which, in turn, can be determined by distance and speed measurements.

Ladder to the Stars

The tale of the baseball-playing robot provides a good way of understanding how the value of the cosmological parameter omega might be experimentally resolved. To determine if the universe will eventually collapse, astrophysicists are attempting to find the deceleration parameter of the cosmos. To achieve this, they are making careful surveys of the distances and recessional velocities of extremely distant galaxies, with hope of pinning down the rate by which the expansion of the universe is slowing down. If this rate is high enough, then omega is greater than one, and universal contraction is inevitable.

Technically speaking, the present-day value of the deceleration parameter, labeled q_0, is defined as a function of the current rate of change per unit time of the Hubble parameter, H. (For the sake of simplicity, I am omitting the mention of some additional factors.) The value ranges, $q_0 < 1/2$, $q_0 = 1/2$, and $q_0 > 1/2$, correspond to open, flat and closed universes respectively. H, in turn, is defined as the rate of change of the universe's scale factor, R, divided by the scale factor itself. The current value of H is called H_0. H_0 is much easier to estimate directly than q_0, because q_0 can only be determined by observing the motion of extremely distant galaxies, those representing an earlier era of the universe. (Because of their great distance, their light takes an exceptionally long time to travel here.)

Although a precise value of the deceleration parameter would be necessary for a *direct* determination of omega (without needing to know anything else), knowledge of H_0, along with an estimate of the density of the universe, would yield a value for omega as well. It can be shown that the critical density, from which omega can be derived, is equal to the Hubble parameter squared, multiplied by a factor of 1.9×10^{-33} grams per cubic centimeter. Therefore, while astronomers pine for the elusive q_0, they realistically pin their hopes on finding H_0 first. The Hubble parameter is usually measured by finding the recession rates of moderately distant galaxies (galaxies outside of the Local Group, but hav-

Galactic Speeding Tickets

ing about the same age as the Milky Way), and dividing by the galaxies' distances to Earth.

Therefore, to determine H and q_0 experimentally it is necessary to construct a detailed map of the distances and velocities of a large sampling of galaxies out to the farthest reaches of the observable universe. These speeds are fairly simple to measure by use of the Doppler effect. Recall that galactic spectra are redshifted at a rate dependent on galactic recessional velocities. By comparing the shifted positions of the spectral lines of known elemental components of a galaxy to their normal spectral positions, the galaxy's speed can be ascertained.

It is far more difficult to figure out the distance to a remote galaxy than to figure out its velocity. There is no cut-and-dried mechanism, such as redshifted light, to provide a simple answer to the question of location. Consequently, the *cosmological distance ladder* (as mentioned in Chapter 3), a rung-by-rung distance hierarchy, has been constructed to aid in this determination.

The first step in this ladder is the method of trigonometric parallax. Parallax is the phenomenon of distant objects appearing to move slightly when viewed from two different vantage points. Place your palm about 10 inches in front of your nose and close one eye and then the other in succession, and you'll experience this phenomenon. By using a bit of trigonometry you could easily use the amount that your palm seems to shift to figure out its distance from your face.

In astronomy, parallax is used to determine the distances of stars that are less than 4000 light-years away. Beyond this scope, the method is no longer accurate. In order to observe the shifting of stars, telescopic readings are taken from opposite points on the Earth's orbit (during winter and summer, say). Once these measurements are compared, a mathematical formula indicates how far away these stars are. The results are then used to calibrate higher rungs on the distance ladder.

The next step, determining the distances to remote stars in our own galaxy, as well as to galaxies that are fairly close to ours, can be achieved to a reasonable degree of accuracy by a number of more sophisticated methods. The most common among these is the Cepheid variable technique, developed by Henrietta Leavitt in 1912 (see Chapter 3 for more details). Recently, this method of using Cepheids as stellar beacons has undergone substantial refinement by use of near-infrared telescopy. Wendy Freedman of the Carnegie Observatories and Barry Madore of Caltech have noted that the Cepheids' near-infrared light passes through interstellar dust much easier than ordinary light, and therefore provides more accurate distance readings.

Supplementing the use of Cepheids as standard candles are methods relying on RR Lyrae variables and planetary nebulae. The RR Lyrae method is quite an old technique that has undergone recent improvement by use of better observational equipment. It was first used by Harlow Shapley around the turn of the century to estimate the Milky Way's size. Through parallax and statistical methods, it has been found that RR Lyrae—variable stars with pulsational periods ranging from a few hours to a day—all have roughly the same absolute magnitude. Thus, once such a short period variable is sighted, its distance can be measured by comparing its apparent brightness to its intrinsic brightness. Until recently, these stars were considered too dim—they're not as bright as Cepheids—to be used for extragalactic distance measurements. However, in 1987, Chris Pritchet and Sidney Van den Bergh used a CCD (charge-coupled device) detector attached to a telescope on Mauna Kea in Hawaii to observe RR Lyrae outside of the Milky Way. They detected these variables in M31, the great spiral in Andromeda, and used them as an independent test of the galaxy's distance, finding it to be 2.5 million light-years away. Since that time, RR Lyrae have been found in other nearby galaxies as well. Undoubtedly, in the coming years, astronomers will increasingly employ RR Lyrae as secondary means of confirming the extragalactic distance scales determined by more standard methods, such as Cepheid variable tests.

Planetary nebulae, the dusty clouds discovered by Herschel and thought to be the remnants of stellar atmospheres, provide yet another way of calibrating spatial yardsticks. Many of these gaseous objects, such as the Ring and Dumbbell nebulae, display very little variation in intrinsic brightness. For this reason, scientists such as George Jacoby of Kitt Peak National Observatory and Robin Ciardullo of Penn State have found these bright objects to be excellent standard candles and have used them to determine distances to other galaxies, even beyond the Local Group.

The Great Debate

In order to establish precise values of the Hubble and deceleration parameters, many astronomers have tried to extend the distance ladder to include the farthest observable galaxies. Most prominently, two teams of scientists, one group led by Allan Sandage of the Mount Wilson Observatory and his Swiss collaborator Gustav Tammann, the other by Gérard de Vaucouleurs of the University of Texas at Austin, have used a barrage of indicators to establish independent methods for determining the current values of H,

Figure 20. Allan Sandage, b. 1926. (Courtesy of AIP Niels Bohr Library, Dorothy Crawford Collection.)

as well as the age of the physical universe. Much to the dismay of the astronomical community, these research groups have found dramatically different estimates of these values.

Over a number of years, starting in the early 1970s, Sandage and Tammann employed a cautious, step-by-step technique to determine the cosmological parameters, relying, for various distance scales, on what they saw as the best available indicators. They began their estimation with what was known at the time from Cepheid variable methods and other extragalactic techniques. Augmenting their data with information about the relationship between galaxy type and absolute luminosity (light output), measurements of the brightness of distant blue and red variable stars, as well as other indicators too numerous to mention, they developed an approximation of the distance to the Virgo cluster, the nearest large group of galaxies. They used this approximation, in turn, to estimate the Hubble parameter.

In 1977, a discovery by Brent Tully of the University of Hawaii and Richard Fisher of the National Radio Astronomy Observatory concerning the spectral properties of galaxies helped Sandage and Tammann, as well as other astronomers, to render their estimates even more precisely. Tully and Fisher found that the width of the 21-centimeter line in the neutral hydrogen spectrum, which provides an indication of how fast a galaxy is rotating, has a strong correlation with the intrinsic brightness of the galaxy. In other words, brighter galaxies spin faster and therefore have broader spectral lines at particular wavelengths.

Using a modified version of this technique, called the Tully–Fisher relation, Sandage and Tammann revised their value for the distance to the Virgo cluster and honed in on an estimate of the Hubble parameter. Their figure, of approximately 50 kilometers per second per megaparsec, with about a 10 percent uncertainty, indicates that the observable universe is about 18 billion years old. (A megaparsec is about three and a quarter million light-years). These estimates of a very old universe easily accommodate other cosmological and geologic scales, such as the age of the solar system and the age of Earth. Also, a low value of H_0 corresponds to a low critical density, making it much more likely that the universe is either closed or flat, rather than open. Most astronomers would accept these values completely if it weren't for the blatantly contradictory results provided by Gérard de Vaucouleurs in 1978. Using a unique combination of indicators, he performed a variety of mathematical analyses to arrive at a figure for the Hubble parameter of 100 kilometers per second per megaparsec, with about a 10 percent uncertainty. This value implies that the universe is only nine billion years old, a bit of a dilemma for those reckoning the evolution of the solar system to take up a significant chunk of that age. If de Vaucouleurs is correct, galaxies, the solar system, and the Earth itself must have evolved from the Big Bang very quickly indeed.

De Vaucouleurs' high value for H_0 also presents complications for closed universe models. It implies an extremely large value for the critical density, almost 2×10^{-29} grams per cubic centimeter. The current estimate for the visible mass density of the universe is much lower than this amount; invisible mass could hardly make up the gap. Therefore, if de Vaucouleurs is correct then the universe would likely be open.

Because there is clearly no overlap between the estimates of de Vaucouleurs and those of Sandage and Tammann, scientists have been at a loss to explain this discrepancy. Obviously, only one of these values could be right, but which one? Both methods seemed to be based on impeccable statistical and astronomical principles, so it was not certain at all what could possibly be amiss.

Moreover, there is a growing minority of scientists who advocate an even lower Hubble parameter value, suggesting that it lies below 40 kilometers per second per megaparsec. In the most extreme case, cosmologist Tom Shanks of the University of Durham, England, has used a revised interpretation of x-ray data from galactic clusters to argue for a value between 20 and 30. This low figure would ensure that the cosmos is ancient enough—at least 20 billion years old—to accommodate even the oldest known stars. It would also virtually guarantee that the universe is closed. However, due to the well-respected, meticulous work of researchers such as Sandage, Tammann, and de Vaucouleurs, most astronomers assign little credibility to values of H_0 below 40.

To make matters even more confusing, a team of researchers, led by Kitt Peak astronomer Michael Pierce, announced in September 1994 that they have employed the Cepheid variable method to determine the precise distance to the Virgo cluster. Using the high-resolution camera on the Canada–France–Hawaii telescope on Mauna Kea, they obtained a distance value that places the Hubble parameter at almost 90. It is too soon to say whether this latest estimate of H_0 is the actual value.

Supernova Speedometers

In the 1990s, improved techniques and more powerful telescopes offer increased hope for enacting a precise determination of the Hubble parameter, and finally resolving the seemingly endless debate about the size and age of the universe. In addition, because of this increased capability, expectations are high that a value can be pinned down for the deceleration parameter as well. Until recently, galactic distance surveys simply didn't press far enough to discern the subtle effects on velocity brought on by q_0. But several new methods, including supernova surveys and gravitational lensing of quasars, offer renewed hope for detecting deceleration and thereby revealing the fate of the cosmos.

Supernovas have long been considered outstanding standard candles, because they can be readily detected in remote galaxies. Researchers, such as Robert Kirshner, of the Harvard astronomy department, have systematically studied the properties of these exploding stars and have developed clever ways of deriving their distances from the sizes of their expanding shells. These methods have added to their utility as significant cosmological bench marks.

Traditionally, supernova bursts were considered rare events, detected only with great serendipity. This infrequency presented a considerable drawback for astronomers, who couldn't just sit around for months waiting for an explosion to occur. Furthermore, low counts meant bad statistics, a problem that could little be afforded in the exact determination of H and q_0.

Recent innovations, however, have improved all this. By using stronger telescopes, scanning larger portions of the sky with more frequency, and employing powerful computers to single out and record large numbers of sightings, supernovas can be spotted and measured more regularly. With hope of finding a precise value for q_0, the British–American collaboration led by Saul Perlmutter, Carl Pennypacker, and Gerson Goldhaber of Berkeley has developed a "supernova factory" at the Isaac Newton telescope at Las Palmas in the Canary Islands. Examining the light from more than 2500 galaxies, they expect to find hundreds of starburst events, yielding extremely precise results.

Hopes that the value of the deceleration parameter is in sight have been buoyed by the Berkeley team's 1992 discovery of the furthest known supernova (detailed in the Introduction). This type-Ia burst has been estimated to have taken place a distance of over five billion light-years away. (Type-Ia supernovas are known for their utility as standard candles; they each have approximately the same intrinsic brightness.) More data of the sort gathered by the Berkeley astronomers will need to be accumulated, however, before an independent estimate of q_0 might be reached.

The Magic of MERLIN

Another promising line of attack on the distance-measurement front stems from a 1979 discovery that massive galaxies tend to distort the light traveling to Earth from distant objects, sometimes resulting in visible multiple images, arcs and rings. This phenomenon, called gravitational lensing, is strongly dependent on how far away the sources are from Earth. Hence, detailed analysis of this occurrence can provide an excellent gauge of distance.

To understand how a gravitational lens works, it is useful to consider the operation of an ordinary lens, such as a magnifying glass. A magnifier serves to bend the paths of distant light rays that travel through it, converging them to a single point known as the focus. Any object placed in front of the lens, such as a tiny insect, must be viewed with convergent light. Consequently, the light rays illuminating the bug appear to emanate from an enlarged image directly under the glass. Thus, in summary, the glassy material of the lens acts to distort our perception by altering the natural straight-line course of light and creating illusory imagery.

The only significant difference between an ordinary and gravitational lens is that, in the latter case, the bending of light rays is achieved through relativistic spatial curvature, instead of by going from one medium to another. Specifically, in gravitational lensing, the geodesics near a

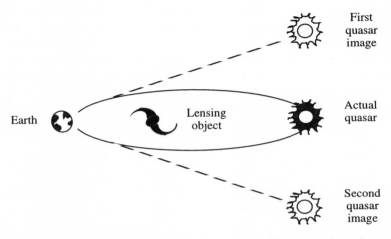

Figure 21. Gravitational lensing of a quasar. The mass of an object (between Earth and a quasar) causes space to warp in its vicinity. Therefore, the light from the quasar, as it heads toward Earth, bends around the intervening body. When terrestrial observers view this light, they trace the rays backward to two imaginary regions, thus seeing a pair of images.

large galaxy or other massive object are transformed by the presence of matter. Therefore, light from a remote part of space, taking the shortest path it can toward Earth, finds itself bifurcated into two of more distinct packets. When these light bundles reach our eye (through a telescope), they appear to have come from several different sources. Therefore, we see multiple images of the source object.

One of the most commonly observed gravitationally lensed objects are the incredibly bright, extraordinarily distant celestial dynamos known as quasistellar objects (quasars). Quasars were discovered during the early 1960s, when radio telescopes detected continuous streams of highly focused radiation, far more than that emitted by any other bodies of comparable size. Quasars are literally hundreds of times brighter than our galaxy, yet they appear to be as small as stars. Because of their intense, pinpoint luminosity, their gravitationally lensed images are relatively easy to find.

Gravitational lenses work well as distance gauges, since quasars tend to vary in brightness periodically. When the flickering beams of light get divided by a lens, two or more images are produced. Since it takes different amounts of time for light to travel along each of these various paths, a time delay becomes built in. Thus, the multiple images don't flicker at the same time. By measuring this lag time, the distance to the quasar can readily be determined—assuming that the mass distribution of the lens is well known. And, by using the velocity readings obtained by the quasar's Doppler shift, a value for the Hubble parameter can be ascertained.

A team headed by Jacqueline Hewitt of MIT, working with Ed Turner of Princeton, has been engaged for several years in an ongoing search for lensed quasars. Their measurements are being taken at the Very Large Array (VLA), a collection of 27 rail-mounted, 80-foot-wide radio telescopes spanning the San Augustin plains of New Mexico. Some of their work has proven quite surprising. An early study of a double image quasar, analyzed by Hewitt and her graduate student Joseph Lehár, had a time lag between flickers of about 540 days. Using calculations based upon an estimate of the mass distribution of the intervening galactic cluster, a Hubble constant of about 40 was determined. This would make the universe over 24 billion years old, a value much greater than all previous assessments. However, because of the considerable guesswork in determining the mass distribution of the cluster, Hewitt is far from certain about this result.

To render her work more precise, Hewitt has been trying to measure systems for which there is less doubt about the mass distribution. For example, she has been looking at the arc-like patterns formed by a lensed quasar that she discovered in 1986, called the Einstein Ring. This particular lens has the advantage that its shape is well known; it is formed by a single galaxy. Thus, its material distribution is far more certain than in Hewitt's earlier case, making analysis much simpler.

Meanwhile, over in Great Britain, researchers from the University of Manchester are performing some magic of their own, using an extremely sensitive detector named MERLIN (Multi-Element Radio-Linked Interferometer Network). Like the VLA, MERLIN is not just a simple telescope, but rather a range of detectors spread out across the English countryside. The effect of such a large distribution is to simulate a telescope approximately 140 miles across.

In a systematic three-year search for the value of the Hubble parameter, Manchester radio astronomers Ian Browne and Peter Wilkinson, together with Tony Readhead of Caltech, have been planning a survey, using MERLIN, of 20,000 radio sources. According to Readhead, they hope to find at least 20 to 40 gravitational lenses, concentrating their analysis on those possessing the same simple characteristics as the Einstein Ring. They are very hopeful that the Hubble parameter will soon be within their grasp.

Hubble Trouble

In April 1990, NASA's space shuttle *Discovery* carried into orbit the Hubble Space Telescope, a device with the

Galactic Speeding Tickets

Figure 22. Jacqueline Hewitt. (Courtesy of Jacqueline Hewitt.)

potential to revolutionize astronomy. By scanning the skies without the hindrance of atmospheric distortion, it has the capacity to reach out farther than any other telescope in history. For this reason, it offers an excellent chance for precise measurements of H_0 and q_0, reliable enough, perhaps, to determine if the universe is open, closed, or flat.

NASA has had a streak of bad luck since the fatal *Challenger* disaster of 1986. Funding has been cut by the U.S. government for many important projects, so morale has been low. It was hoped, therefore, that the Hubble telescope would provide a boost of confidence for the space program. Instead, massive technical problems with the telescope threatened to turn NASA into a virtual laughingstock.

Before the Hubble was launched, it was equipped with a 96-inch-wide main mirror, the smoothest ever made. The Connecticut factory that ground this mirror took pride in its apparent lack of flaws. Unfortunately, once the telescope was already set into space, NASA scientists discovered that the mirror had an improper curvature. The Hubble was diagnosed as having a case of myopia. Incoming light failed to focus properly after being bounced off the misshapen mirror. Consequently, fully half of the experiments proposed for the Hubble needed to be postponed or scrapped.

A pessimist views a glass and says it is half empty; an optimist, half full. In spite of the 50 percent cut in projects, some important astronomical work took place using the flawed Hubble, including detailed measurements of cosmological parameters. Computers helped to reduce the distortion problem by making appropriate adjustments. So all in all, astronomers weren't as disappointed as they might have been.

Allan Sandage, for one, was quite happy with results that he obtained from the space telescope during its flawed period, confirming his earlier predictions that the universe is large and old. In the summer of 1992, he recalibrated the distances of a large number of type-Ia supernovas and obtained a value of 45 for the Hubble parameter. After discovering a minor error, he later revised this estimate to be 51, consistent with his 1970s findings. He undertook this exhaustive survey with a number of notable astronomers, including Nino Panagia, Abhijit Saha, and Duccio Machetto, of the Space Science Institute in Baltimore (the base of operations for the Hubble), and his old collaborator Gustav Tammann of the University of Basel, Switzerland.

The scheme they devised involved measuring the distance to one of the Milky Way's neighbors, the IC 4182 spiral dwarf galaxy. In 1937, this galaxy was host to a famous type-Ia supernova burst, the apparent brightness of which was measured, at the time, by Fritz Zwicky. If the distance to the galaxy could be measured, then its intrinsic luminosity could be found. Armed with the value of this prototypical absolute brightness, Sandage and his collaborators reasoned, one could then measure the distances to other, far more remote type-Ia supernovas.

To assess the distance to IC 4182, the team decided to record the light output of 27 of its Cepheid variables, most too dim to observe from Earth-bound telescopes. To view these Cepheids without atmospheric interference, they obtained permission to use the Hubble's Wide Field and Planetary Camera for a 47-day period. To measure all incident photons with high precision, this camera's field of view is covered by a 1600×1600 array of CCD (charge-coupled device) pixels. Because each CCD, at any given time, can record the impact of a single light particle, the combined array of over two million of these devices insures a steady stream of optical information from space.

The results from the space telescope were superb—extremely clear in spite of the main mirror's optical flaw. Elaborate computer software scanned through the CCD data and sorted out the light signals from the noise, revealing a host of useful data. Detailed measurements of the Cepheids' pulses provided enough information to determine accurately the distance to IC 4182; it is 4.94 megaparsecs away. Thus, the 1937 supernova burst must have taken place about 16 million light years away.

Once Sandage and his co-workers discovered how far away the burst was, they used this fact to calibrate the absolute luminosities of all other known type-Ia supernovas. This information, coupled with the apparent brightness data collected over the years, yielded the distances to each of the galaxies housing these exploding stars. Finally, the team plotted a distance versus velocity chart for these galaxies and obtained an estimate of the Hubble parameter.

In 1993, American astronauts installed long-awaited optical corrections to the space telescope's flawed main mirror, enabling astronomers to peer much farther across the void. This repair mission was a great technical success for NASA, requiring a well-trained crew to spend extended periods of time walking in space. There were five space walks, totaling more than 30 hours, longer than ever before in history. The 300 million dollar device, called the corrective optics space telescope axial replacement (COSTAR), that was put in to correct for the myopia needed to be placed in front of the mirror by use of computer-controlled automated arms. New solar panels were installed and failed gyroscopes, which keep the telescope stable in orbit, were replaced. At last, thanks to the intrepid crew of the space shuttle *Endeavour*, the Hubble no longer was near-sighted and could carry out its mission for eager astronomers on Earth.

The repaired Hubble now has the incredible capability to gaze toward the very edge of the observable universe and view stars, galaxies, and quasars as they were forming 14 billion years ago. It is performing remarkably close to original expectations, considering its earlier problems. In some

Galactic Speeding Tickets

ways, it is working even better than originally anticipated. According to David Leckrone, a senior scientist at NASA's Goddard Space Flight Center in Greenbelt, Maryland, "The bandages came off ahead of schedule, and our patient has incredible clarity." Ed Weiler, another NASA scientist, is equally enthusiastic: "It is fixed beyond our wildest expectations."[1]

By the end of this decade, there is an outstanding chance that we'll learn the values of the long-sought deceleration and Hubble parameters. Knowledge of the deceleration parameter, the hardest of the two to obtain, would immediately reveal our cosmic fate. But if we still can't pin down q_0, then obtaining a precise value of the Hubble parameter, along with information about the mass of the cosmos, would do just as well.

In the next chapter, we'll examine the latter half of this cosmological program: weighing the universe and determining its material content. As we'll see, to do this would require astronomical techniques that are just as bold and far-reaching as those we've already discussed, utilizing the new space-based light-recording devices such as the Cosmic Background Explorer and the Hubble Space Telescope, as well as other advanced telescopes, based on land, that are almost as powerful. Such an ambitious venture to resolve our fate would entail solving a deep mystery of the cosmos, one that has baffled astronomers for the past few decades, the problem of missing mass.

THE SEARCH FOR MISSING MATTER

For the first time there is evidence of enough dark matter to support the idea that the universe is closed.
-Richard Mushotzky, Goddard Space Flight Center

Weighing the Universe

The best hope for determining the destiny of the universe, and to learn whether or not our distant descendants will be pulverized in a Big Crunch, involves a fairly straightforward method. First, measure the current value of the Hubble parameter. Sandage, Tammann, and others are working on this; odds are that it is between 40 and 75. Second, square this figure and multiply by the quantity of 1.9×10^{-33} grams per cubic centimeter (this is an approximation) to find the critical density. The larger the Hubble parameter and the faster the expansion of the cosmos, the greater the density needed to arrest the expansion. Assuming a low value for H_0, then the critical density is about 5×10^{-30} grams per cubic centimeter. Third, divide the critical density into the actual material density of the cosmos to get omega. Finally, decide if omega is less than one, equal to one, or greater than one.

The original estimates for the material density of the universe were all much lower than the critical density. This is because they were based on the erroneous assumption that most of the mass in the cosmos is luminous. One can easily understand why this supposition was made; it is true for relatively small regions such as our solar system. The sun, a luminous body, embodies well over 99 percent of the material content of the solar system. Consequently, in the sun's domain, the ratio of luminous matter (the sun) to dark matter (the planets and asteroids orbiting the sun) is a thousand to one. Extrapolating this ratio to the rest of the cosmos, one arrives at a very low figure for its overall density.

A good gauge of the amount of visible matter in space is the mass-to-luminosity ratio (M/L). The total amount of light produced by the bright objects in a particular region of space is fairly easy to measure using, for example, CCD-equipped telescopes. Multiplying this quantity by the M/L ratio yields the amount of mass in that region. Given the amount of detectable radiation in space, it turns out that a M/L value of at least 1000 overall would be needed for the universe to recollapse. (The precise value needed depends on the Hubble parameter.)

Employing the techniques of nuclear astrophysics to analyze the processes that led to the creation of the least heavy (non-hydrogen) elements, the universe's baryon-based M/L value can be computed. According to the Big Bang model, during each stage of nucleosynthesis, from hydrogen to deuterium to helium to lithium, enormous quantities of baryons and photons were produced in precise ratios. Extrapolating forward in time, the current ratio of baryons to photons is considered to be approximately one part per two billion. (This range is likely to be sharpened considerably in the next few years by detailed analysis of the COBE results.) Estimating the mass of these baryons and the luminosity of these photons, one arrives at a M/L figure of about 70.

Baryon matter—mainly protons and neutrons—is the stuff from which stars and planets are built, and would seem to be the leading contender for the bulkiest substance in the cosmos. Everywhere we look we see stars. Stellar matter is largely baryonic; therefore, one might guess that the universe is mostly baryonic. Making this assumption, one is forced to conclude that the cosmos is open, with a density of less than 10 percent of its critical value. There clearly isn't enough baryonic matter to close it.

Moreover, astronomers cannot account for the vast majority of these baryons. Star and galaxy counts yield a much lower figure than theory would indicate for the cosmic mass-to-luminosity ratio. These experimental results derive from telescopic searches, as well as models of stellar light production. Scientists have been forced to conclude that a large portion of the material content of the universe is invisible.

Hidden Planets

There have been a number of proposed explanations for the missing baryons. Some have postulated the existence of an overwhelming number of "brown dwarfs," stars that don't shine (or shine very little) because they aren't massive enough to engage in nuclear fusion, and "jupiters," planets, like Jupiter, composed of a considerable amount of matter. Because these objects would be far smaller than most shining stars, to make up the missing baryon content of the universe there would need to be thousands upon thousands of nonburning stars and bulky planets for each visible orb. Efforts are underway to try to detect and count these dim celestial bodies which have been impossible to spot, until recently, because of the floodlight illumination of neighboring suns.

A number of recent innovations in astronomical and computational technology have aided in the search for brown dwarfs and jupiters. The Hubble Space Telescope has been equipped with special filters to block out the illumination from the central disk of a star while permitting the peripheral light to pass through. Scientists hope to use these filters to help them detect the faint reflected light from large planets. A new sort of observatory, the Infrared Astronomical Satellite (IRAS), has been programmed to collect and analyze a flood of low-frequency interstellar radiation with the aim of uncovering new phenomena. Finally, new radio-astronomical techniques have been designed to detect even minor fluctuations in radiative emissions, such as those likely to be produced by interfering planets.

In July 1991, the British team of Andrew Lyne, Matthew Bailes, and Setnam Shemar reported the discovery of a Jupiter-sized planet orbiting a distant pulsar. A pulsar is an extremely dense, fast-spinning neutron star, about a dozen miles in diameter. It gives off radiation in regular pulses, an effect caused by its rapid rotation.

Although serious problems were found with the British team's report—the data were found to be statistically insignificant—it was soon followed by accounts of far more promise. At the January 1992 meeting of the American Astronomical Society, Alexander Wolszczan of Penn State University gave a talk about two, and possibly three, planets orbiting a different pulsar, labeled PSR 1257+12, that he and Dale Frail, of the National Radio Astronomical Observatory in Green Bank, West Virginia, had discovered. They had detected the planets in 1991 using a 1000-foot-diameter radio telescope at the Arecibo Observatory in Puerto Rico. The planets are located more than seven thousand trillion miles from Earth in the constellation Virgo.

At the time of their 1992 report, Wolszczan and Frail considered their results to be statistically incomplete and began a search for more evidence to support their discovery. Finally, in the April 1994 issue of *Science*, Wolszczan announced success. He presented unequivocal proof that objects of planetary size exist outside our solar system.

The team's detailed verification took a great deal of patience to complete. For almost three years, Wolszczan and Frail mapped out the radio signal emitted by the PSR 1257+12 pulsar, looking for unusual variations that would confirm the presence of orbiting planets. Because the pulses from pulsars are ordinarily very precise, resembling the regular ticking of an atomic clock, the researchers found it possible to detect the gravitational influence of planets on the pulsar by looking at tiny "bumps" in the signal. After months of analyzing such bumps, Wolszczan and Frail concluded that at least three planets orbited the pulsar.

The first planet appeared to be about 3.4 times Earth's mass, orbiting the pulsar every 66.6 days. The second was estimated to be about 2.8 times the mass of the Earth, orbiting once every 98.2 days. Finally, the third was found to be similar in mass to Earth's moon, revolving around the pulsar once every 25 days. Of course, far more extrasolar planets than these would need to be found to account for the missing baryonic mass in the universe.

Another proposed bearer of undetected baryonic stuff is the mysterious end product of massive stars, the black hole. Black holes are so dense that, even if they made up only a small percentage of the universe's volume, they still could comprise a large portion of its mass. Currently, so little is known about them that it is hard to say if they'll provide any part of the answer to the missing baryonic mass dilemma. According to researchers at the Space Telescope Institute, one of the top goals for the repaired Hubble Space Telescope is to survey the sky for signs of these compact relics. With its ability to focus on the centers of galaxies, it is hoped that the Hubble will find evidence there of the characteristic radiation produced from stars being pulled into black hole vortices. If the mass of these objects could then be ascertained, we would have a greater understanding of their roles as reservoirs of large quantities of baryonic matter.

Stellar Carousels

A host of astronomical measurements have presented us with an issue that is even more perplexing than the problem of missing baryons; namely, there is substantial evidence that most of the matter in the cosmos is non-baryonic in nature. This result comes from estimating the mass of the universe, as indicated by the gravitationally induced motions of stars, galaxies, and other objects, and comparing this value to its maximum estimated baryon content. The former vastly overwhelms the latter, demonstrating that space contains far more than just the ordinary elements.

It has been known for a number of years that Newtonian dynamics, applied to objects of galactic size, indicates that the universe's density must be greater than what is

Missing Matter

actually observed. In the early 1930s, Dutch astronomer Jan Oort observed the motion of stars moving above and below the plane of our galaxy and found that these dynamics would require three times as much mass in the stars' vicinities as actually observed.

Oort's study was based on a sort of merry-go-round model of the Milky Way. Like the horses on a carousel, stars enact two distinct types of motion in our spiral galaxy. The first kind of behavior is to orbit the galactic center, and the second is to bob up and down relative to the galactic plane. Just as the dual movements of a carousel's stallions are linked to the strength and speed of its driving engines, the twin behaviors of our galaxy's stars are connected to the power of its gravitational dynamo. This galactic generator is fueled by the matter in the Milky Way. Hence, by studying stellar motions, the mass distribution might be found.

Scientists in the 1930s were not overly concerned by Oort's findings. There was little known at that time about interstellar gases, but it was assumed that there must be other substances besides stellar material in the galaxy. Therefore, astronomers correctly supposed that much of Oort's "missing mass" would turn out to be the matter between stars. Indeed, since Oort's discovery, fully two-thirds of this formerly unseen matter has been detected in the form of cold clouds of gas and dust.

Another set of findings in the 1930s, recorded by Swiss astronomer Fritz Zwicky, presented far more of a problem. Zwicky, working at Caltech, measured the relative redshifts of numerous galaxies in the Coma Cluster, and found that they varied a great deal. This variation could imply only that these galaxies were moving quite rapidly within the framework of the cluster. Yet the cluster hadn't fallen apart; therefore, some central force, presumably gravity, must be holding it together. Zwicky calculated the amount of mass needed to produce this gravitational binding, and found it to be staggering—dozens of times greater than the observable quantity of star matter. The implication was that a significant part of the Coma Cluster is composed of invisible substances.

It wasn't until the 1970s, however, that the missing mass issue achieved recognition as a fundamental problem in cosmology. By then, observational methods had become so effective that it was felt that most of the detectable matter in the vicinity of the Milky Way had already been seen. Therefore, the existence of gaps in the galactic neighborhood's mass distribution, as determined by gravitational dynamics, was perceived as a real problem that needed to be satisfactorily addressed. And, in a bizarre situation no one could suitably explain, one of the biggest of the sky's matter gaps seemed to be drawing our galaxy toward it at incredible speeds.

In 1976, Vera Rubin and several colleagues at the Carnegie Institute discovered this invisible "Great Attractor" in the sky, while performing a study of the motion of our galaxy relative to space. The team analyzed the redshifts of a large (100-Megaparsec radius) spherical shell of spiral galaxies, centered on our own, and found that the Milky Way is moving at a rate of 600 kilometers per second relative to this shell. One should not confuse this figure with the universal Hubble expansion; that isotropic effect has already been filtered out of the equation. On the contrary, this rapid motion is unidirectional and therefore must be due to the gravitational force of a massive hidden object or set of objects.

The direction of this tremendous tug is quite curious. It seems concentrated at a region of space in which there are few apparent galaxies, certainly not enough to supply such a driving force. Apparently, huge quantities of dark matter, detectable only by their gravitational effects, reside in what was formerly thought of as the intergalactic void.

Infrared sky surveys have confirmed the existence of this Great Attractor; its origins and composition, however, remain shrouded in mystery. The IRAS probe has indicated that not only is our own galaxy heading toward this massive body, but other galaxies are veered in its direction as well. On the basis of large-scale infrared maps, Michael Rowan-Robinson of the University of London has estimated the quantity of dark matter in our section of the universe needed to cause such behavior. He has determined that the density of matter in the universe, including this hidden material, is roughly on par with the critical density. In other words, according to this well-regarded survey, omega is at least one, and the universe is either flat or closed. Because there clearly isn't enough baryonic matter in the universe to close it, there must be other massive substances out there causing this effect. Hence, not only is there a missing baryonic mass mystery, there is also an invisible nonbaryonic mass dilemma.

MACHO Men

Assuming that there is enough dark matter to close the universe, or at least enough to generate the gravitational effects detected by astronomers such as Oort, Zwicky, Rubin, and Rowan-Robinson, it is important to determine exactly where it is located. Only then can scientists construct viable theories as to its origin and content. Intuitively, one might imagine several distinct possibilities for its distribution, including, at one extreme, uniform dispersion through space and, at the other extreme, location only within galaxies themselves. Analyses so far seem to indicate that the actual arrangement of dark matter follows neither extreme; most is harbored by galaxies, but large quantities are positioned elsewhere in the cosmos.

Most surveys of the dark matter distribution within galaxies rely on studies of the rotational velocities of stars in their peripheries. Newton's laws, as applied to orbiting objects, mandate that their speeds reflect the amount and organization of mass nearby. The simplest example of this involves the orbital motions of the planets within our solar system. Because the vast majority of the solar system's material lies within the sun, a highly concentrated distribution—planetary velocities tend to taper off with radial distance. Thus Pluto's speed is much slower than that of Venus or Mercury, by a predictable amount. These velocity relations are known as Kepler's laws.

If all of the mass in the Milky Way were distributed according to its *visible* allotment, then a system similar to Kepler's relations would apply to the stars in our galaxy as well. The visible distribution of galactic matter is highly concentrated; the bulk of it is located within a thin, disk-shaped region that surrounds a central "bulge." Because of this saucerlike shape, one would expect that stars closer to the galactic plane (and bulge) would experience a stronger gravitational pull, and therefore orbit the center of the Milky Way more quickly than stars farther away.

The actual behavior of stars in our galaxy is quite different than one would predict. If one plots their speeds versus their distances from the center one sees virtually no drop-off whatsoever. These "flat rotation curves" signify that the galactic mass distribution is much less centrally concentrated than the visibly disklike structure would indicate. Rather than being confined to the central "saucer," it seems that much of the matter in the galaxy is scattered throughout the periphery, at considerable distances above and below the plane.

In order to determine how far out the Milky Way's dark matter extends, a team of astronomers led by Douglas Lin of the University of Santa Cruz set out to measure the motion of 250 stars in the Large Magellanic Cloud galaxy (a neighbor of the Milky Way). By analyzing the slight changes in the movements of these stars over the period from 1974 to 1989, they hoped to detect the long-range gravitational influence of our galaxy, and thus determine its mass distribution. Utilizing readings taken by the four-meter telescope at the Cerro Tololo Inter-American Observatory in the Chilean Andes, they looked at motion so small and so far away that it has been described as watching grass grow in New York from London.

Their results, presented at a meeting of the American Astronomical Society in June 1993, were absolutely startling. They found that the Milky Way is at least five times as wide as once believed, encompassing the Magellanic Clouds and other galactic satellites within its remote boundaries. During each orbit the Large Magellanic Cloud loses chunks of matter to our galaxy. This debris is evident to astronomers as a cometlike trail following the Cloud. According to Lin, "We're witnessing the ongoing process of galactic cannibalism." Eventually, in about 20 billion years, the Large Magellanic Cloud will be completely broken up and absorbed, forming part of our own spiral realm. As a result, the Milky Way will seem much fatter.

Filling the vast domain between the central disk and the nether reaches of our galaxy is a mysterious halo of invisible material. The interactions of this substance with the

Large Magellanic Cloud are strong enough to prevent the Cloud from flying off into deep space. Based on estimates of the dark matter needed to create these tremendous gravitational forces, the Milky Way is at least 10 times as heavy as its stars and gas would suggest. Thus over 90 percent of our galaxy is composed of objects that cannot be seen by telescope.

There have been numerous theories about the composition of this dark "halo matter." Some researchers suggest that it is spread out across the galactic periphery in the form of innumerable subatomic particles that interact weakly with normal matter. These so-called WIMPS (weakly interacting massive particles) would have the theoretical advantage of being nonbaryonic in nature and therefore, their presence wouldn't violate the strict limits on the baryon content of the universe imposed by Big Bang nucleosynthesis models.

Other scientists contend that a large part of the Milky Way halo's missing matter is hidden in the form of either brown dwarfs, black holes, or jupiters, as noted earlier. Generically, these objects have come to be called MACHOS, for massive compact halo objects. Because they would be exclusively composed of baryons, and there are limits to the baryonic content of the universe, MACHOS couldn't comprise all of the dark matter. However, there is recent evidence that at least some portion of the galactic halo's material is made up of MACHOS.

In September 1993, two independent teams of astronomers, the first group from Australia and the United States, and the second from France, reported at conferences in Capri and Gran Sasso Italy the first direct detection of a massive halo object. The American–Australian group, consisting of 16 scientists from five institutions, including the

University of California at Berkeley, gathered their results from a detailed survey of more than 11,000 images, obtained from the 50-inch telescope at the Mount Stromlo observatory in Australia. The equipment that they used included one of the largest CCD cameras used in astronomy. Analyzing the light output of 1.8 million stars, they noticed a single incident in which one of the stars—possibly a red giant—brightened and then dimmed over a period of approximately two months. They attributed this symmetric rise and fall of light emissions to the focusing properties of an intervening MACHO, acting as a gravitational lens. The change in brightness was believed to be caused by the dark object's bending of the star's light rays before they reached Earth. The French team, based at Saclay, noted similar behavior in two other stellar sources. The mass of each of these MACHO candidates was estimated to be about one-tenth that of the sun, roughly the size expected for a brown dwarf. Because only a few of these objects have been catalogued so far, a search is underway to supplement these rare events with more data.

Intergalactic Shadows

Meanwhile, as researchers struggle to detect dark matter in our own galaxy, other astronomers engage in a quest for missing mass scattered throughout the intergalactic void. If, as many theorists believe, omega is greater than or equal to one, and the universe is closed or flat, then there must be an enormous quantity of hidden mass distributed throughout space, including places where there is no star formation.

To probe the nature of intergalactic dark matter, a team of astrophysicists, led by Anthony Tyson of Bell Labs in New

Jersey, as well as experimentalists from Princeton, Cambridge, and the University of Arizona, has embarked on a multiyear project to look for gravitationally lensed images of very distant galaxies. Employing ultrasensitive CCD detectors at the Cerro Tololo Inter-American Observatory telescope, they've scanned the skies for millions of faint blue galaxies, which they have used as the background to explore lensing effects. As a signal for the presence of dark matter, acting as gravitational lenses, they've looked for examples of distorted galactic images. Once these stretched-out pictures are observed, computer programs have been used to reconstruct the mass distribution that would have caused such distortion. So far, their results have been inconclusive, but they are hopeful that they will soon unveil large quantities of dark matter.

Another search project for hidden mass that has received quite a lot of attention involves a study of the behavior of a huge cloud of hot, fast-moving gas, located in the NGC 2300 galactic group. The NGC 2300 group consists of three small galaxies located about 150 million light-years from Earth in the direction of the constellation Cepheus. Utilizing the international ROSAT (Roentgen Satellite) x-ray observatory, a group of astronomers measured the gravitational stability of this cloud, reasoning that because there wasn't enough visible mass to hold it together, there must be a considerable amount of dark matter present. Their hunch paid off. Computer analysis of the amount of matter needed to stabilize the cloud revealed a bounty of unseen material.

After calculating the mass of this dark matter to be 25 times greater than that of visible material in the vicinity, team scientists John S. Mulchaey of the University of Maryland, Richard F. Mushotzky of NASA's Goddard Space

Flight Center in Greenbelt, Maryland, and David Burstein of Arizona State University, made a gleeful announcement at the January 1993 meeting of the American Astronomical Society that—in the words of Mushotzky—they had found strong evidence to "indicate that there may be enough material to close the universe."

According to Mulchaey, speaking to reporters at the meeting, "The new findings add much weight to the theory that most of the mass of the universe consists of dark matter, the precise nature of which remains unknown to scientists [The discovery] is the first reliable observation of dark matter in 'poor' [small galactic community] regions of space."[1]

The importance of the ROSAT findings, if confirmed in other parts of space, cannot be overstated. If the underpopulated regions of the universe, such as small groups of galaxies, are full of invisible matter in about the same ratio detected by the ROSAT team, and the Hubble parameter is low enough (less than 60, let's say), then omega is likely greater than one. In that case, theory indicates that the density of the universe would be large enough to cause cosmic collapse to take place someday. X-ray surveys of other gas clouds will help to sharpen these results, and indisputably determine whether or not the cosmos is indeed closed.

In a bit of cosmological controversy, scientists Mark Henriksen of the University of Alabama and Gary Mamon of the Observatoire de Paris-Meudon in France have recently asserted that the high-mass estimates of the NGC 2300 region were based on faulty analysis. In a research article published in the February 1994 issue of the prestigious *Astrophysical Journal Letters*, they presented a calculation showing that the invisible matter detected in NGC 2300 is only one-fifth as massive as the claims of its discoverers would indicate.

When I spoke to Mulchaey about the apparent discrepancy, he responded that there is no fundamental contradiction between his own team's work and the more recent results by Henriksen and Mamon. Moreover, he remarked that there is no animosity between the two groups of researchers. Mulchaey said, "Mark [Henriksen] and I are friends. He put in the wrong numbers to get his results. He used our old data, not our new data. Our main result is still correct, that these groups [of galaxies] are dominated by dark matter."

Mulchaey indicated that much more work needs to be done to obtain a credible value of omega. He remarked, "[Currently] there are many uncertainties—the temperature and density as a function of radius, for instance. We are in the process of looking at many more systems, investigating them at rapid pace."[2]

The ROSAT experiment is now over; it has literally run out of gas (an x-ray-sensitive gas used by the detector has been depleted, rendering the device inoperable). Scientists hope that investigations to follow will provide us with precise information about the dark matter content of deep space, including regions relatively free of clusters. Much is at stake. If, on the one hand, dark matter is largely confined to galaxies and clusters, then the cosmos is probably open. If, on the other hand, the high ratio of dark-to-luminous material found by Mulchaey, Mushotzky, Burstein, and their collaborators is true even for domains once believed to be scarcely populated, then the universe is likely closed. In that case, it is destined to retract someday in a Big Crunch, and perhaps even undergo future cycles of oscillation.

Missing Matter

Running Hot and Cold

In the past decade, there has been a great deal of speculation about the composition of the *nonbaryonic* invisible material that many scientists believe exists in large enough amounts to close the universe. In the cosmological pageantry of dark matter candidates that has taken place, a number of once-viable contenders have been brought up to the stage, and paraded before scientists, had their virtues and blemishes scrutinized, and been summarily dismissed. Theoretical astrophysicists, once reasonably confident that the nature of the missing mass would be easily resolved, now scratch their heads and wonder how to make sense of this mystery.

There are two general categories of nonbaryonic dark matter: hot dark matter (HDM) and cold dark matter (CDM). HDM mainly refers to a candidate that was once considered extremely promising, the massive neutrino. Until the 1970s, physicists generally assumed that the neutrino, an extremely common, high-speed particle, was absolutely massless. However, those trying to solve the missing matter dilemma soon realized that even if neutrinos possessed tiny masses they are so common that this mass would dominate the universe. Quite likely, their gravitational force would prove enough to close the cosmos; hence they provided a natural solution to the dark matter question. The race was on to try to determine the mass of the neutrino.

In 1980, much to the delight of theorists, a team of Russian experimentalists led by V. A. Lyubimov claimed success. They measured a neutrino mass that would be sufficient for such particles to dominate the present-day density of the cosmos. Unfortunately, however, subsequent studies have failed to reproduce this result. Today, scientists believe that if the neutrino did have mass it would be too small to enable us to account for all of the missing matter.

HDM models, including massive neutrino schemes, have fallen out of vogue for another significant reason, having to do with the problem of structure formation in the early universe. Neutrinos, and other energetic, extremely fast-moving particles, would serve, in the first moments of the universe, to damp out structure smaller than galactic clusters. The rapid thermal activity of these minuscule bodies would smooth out all perturbations below a particular size, suppressing the formation of individual galaxies. Therefore, if massive neutrinos had dominated the Big Bang, the universe today would look very different, with few galaxies of the Milky Way's size and composition. Because of these difficulties, most researchers have turned to pure CDM theories and mixed CDM/HDM approaches. Pure CDM models were popular during the 1980s, because they seemed to describe structure formation in the universe quite handily. However, as new information about large-scale structure in the cosmos became available, researchers realized that CDM theories fell somewhat short in their predictions. New hybrid schemes were formulated, involving enough CDM to account for galaxy formation and enough HDM to allow for the creation of more extensive configurations.

Theoretical examples of CDM candidates abound, including elementary particles called photinos, gravitinos, and axions, and a mysterious substance called shadow matter. The trouble is, these are all purely hypothetical. Photinos and gravitinos stem from a theory called *supersymmetry*, which requires each boson (integer spin particle) to have a fermion (half-integer spin particle) companion, and vice-versa. Supersymmetry is a theoretical balance between the two major particle spin designations, somewhat analogous to the symmetry, in magnetism, between north and south poles. If supersymmetry is valid, then hypothetical fermions, such as photinos and gravitinos, would exist as counterparts to photons and gravitons (gravity particles), which are bosons. Experimentalists are engaged in an ongoing search for these "supersymmetric companions." So far nothing substantial has turned up.

Axions and shadow matter derive from other high-energy models, too intricate to discuss here in detail. Shadow matter would have the peculiar property that it would interact with normal matter by means of gravity exclusively; no other associations with ordinary substances would be possible. Until a powerful enough accelerator is built, we won't know if any of these postulated particles truly exist.

In trying to resolve the missing mass dilemma, without resorting to hypothetical entities, some scientists have even revived Einstein's old notion of a cosmological constant which he introduced, but quickly discarded, when general relativity was formulated. With a nonzero value of this parameter, omega could be less than one and the universe still be closed. Because the reintroduction of the cosmological constant would mean a modification of Einstein's elegantly simple equation, most scientists still hold out hope that another way will be found to settle the dark matter predicament.

The Closing Circle

The universal serpent readies itself. Does it have the strength and determination to coil, or will it just fall flat? Scientists carefully examine its shape, record its slithering actions, even try to measure its mass, but still haven't developed the power to anticipate its next move. Clever creature it is.

What is the consensus about whether or not the universe will eventually "recoil" back to a point? There is none at present. Many theoretical physicists, for a variety of reasons, favor a closed or flat model of the cosmos. Based on *current* estimates, however, the majority of observational astronomers don't believe that there is enough mass in the universe for it to recollapse.

The search has really just begun for the invisible material that would, perhaps, close the universe. And with each discovery of another pocket of dark matter, the ancients' image of the cosmos as *ourobouros* comes closer to realization. What would the authors of the Hindu Vedic scriptures and the creators of the Egyptian hieroglyphic writings have thought if they knew that their coiled models of time would someday be tested by multibillion dollar atom smashers and massive orbiting space telescopes?

Although there isn't enough evidence, at present, to render a verdict on whether the universe is open, flat, or closed, it is fascinating to imagine what might happen if the universe were to collapse. In the next few chapters, I'll offer a speculative account of what a Big Crunch universal demise, or a cyclical cosmic "resurrection," might be like.

PART 4

UNRAVELING THE CYCLICAL SERPENT

REVERSE PERFORMANCE

> *For this was the Dark—the Dark and the Cold and the Doom. The bright walls of the universe were shattered and their black fragments were falling down to crush and squeeze and obliterate him In the instant, the awful splendor of the indifferent Stars leaped nearer.*
> *-Isaac Asimov, Nightfall*

The Sky Is Falling

It is the year five billion A.D. and the sky is falling. Electronic newspaper headlines have just proclaimed that the universe has stopped expanding and begun contracting. Sirens scream through the air as mobs of panicked earthlings foolishly scamper into the little-used underground shelters, relics of the last interspace war and of no use now. Others blast off in massive space arks, full of thousands of terrified emigrants from our besieged world. But it is to no avail, because the fate that awaits the Earth looms over the entire observable universe as well.

Meanwhile, drawn-out distress cries blare over emergency communication channels, announcing the collapse of the cosmos to the other member civilizations of the intergalactic confederation. But scientists from the other planets know their ungodly destiny already. They have already detected the dozens of blueshifted light signals, seemingly flowing from the faces of distant galaxies from which redshifted spectra used to emanate.

True, the farthest galaxies are still churning out red Doppler-shifted light. But light takes time to propagate and these stellar spirals have not yet had time to broadcast their message of doom. In the intervening millions of years between the times their signals were sent and the present, these galaxies have likely begun to shift their courses and come plummeting in toward our own. It is only a matter of time before the entire sky assumes a chilling, crushing blue hue—at night, as well as by day.

Reporters gather in the Betelgeuse Canteen to ponder how to convey the steady stream of bad news to a weary public eager for signs of relief. Edward R. Murrow, distant descendent of the famous 20th-century journalist, wails, "this is far, far worse than even the Hindenberg disaster." Star reporter Orson Welles, a descendent of the actor, adds, "the crowds are even more frightened than during the first Martian invasion." The Ted Koppel and Dan Rather androids shake their heads in despair as they sip their chock-full-of-radon coffee. Then, as the lights of the cafe dim, Walter Cronkite, the hundred millionth, sighs as he prepares his final news report: "The universe is contracting down to a point, folks, and that is the way it is."

Confronting Collapse

I have just painted a rather far-fetched science fiction scenario—the B-movie version of what might happen if the universal Hubble expansion were to reverse itself. Fortunately, in complete contrast to this bleak picture of a universal demise, chances are that, according to the latest cosmological theories, the dawn of such a cosmic collapse would be quite uneventful for those on Earth (assuming that the Earth still existed). There is really no cause for alarm—not yet, anyway.

There is a long tradition of superstitious panic surrounding any sort of significant celestial alignment or cataclysm, whether real or perceived. As discussed, every 52 years, at the end of one of their calendar cycles, the members of the Aztec civilization would leave their homes and cities, fearing that the end of the world was near. Other ancient cultures greeted the arrival of a comet or supernova with emotions ranging from deep apprehension to sheer terror.

In modern times, the coming of the last days of centuries has commanded similar anxiety. Quite likely, in the same way that apocalyptic terror heralded the beginning of the present millennium, the arrival of the year 2000 will stimulate a rash of prophetic movements offering up their own sinister revelations. And all this pandemonium, if it were to happen, would be triggered by a purely man-made construct, not even a natural catastrophe.

Isaac Asimov, in his short story "Nightfall" (which many critics consider his finest), describes a fictional alien world, Lagash, that experiences this sort of reactive chaos on a cyclical basis. Lagash, situated in the vicinity of six suns, is bathed in light virtually all of the time. Therefore, the race of beings on this planet is terrified of darkness and unaware—because it is always daytime—of the existence of outer space. It is generally believed on Lagash that there are only six suns in the entire universe; there is no cognizance of the billions of other stars in space.

Every 2049 years, however, complete catastrophe devastates the fragile planetary civilization when a special alignment of five of the suns, coupled with a total eclipse of the remaining sun, Beta, allows the sky to go completely dark. Naturally, the darkening of the heavens causes myriads of stars in the sky to appear, unobscured by the shining of the suns. Unfortunately for Lagashian culture, a pitch-black dome speckled with countless points of light causes widespread panic when the citizens of Lagash realize for themselves the cold, dark vastness of space. As a result, each time this particular celestial phenomenon occurs, millions become insane, riot, and then burn down all of the cities, completely destroying all vestiges of society. Thus cyclically, after these turbulent episodes, all civilized life on Lagash is forced to sort through the rubble of what is lost, pick up the pieces, and begin again.

A mysterious cult has developed to preserve the memory of these periodic conflagrations. This group, named appropriately enough, the Cultists, draws its knowledge from reports made by those who survive the upheavals with their sanity intact. These survivors include the drunkards, the mentally deficient, and young children—all groups that don't know enough to be afraid of the night sky. Through this reservoir of knowledge, a Book of Revelations has been composed, documenting the events of each worldwide catastrophe to be passed down to future generations by the Cultists. But because the sources for this text are so poor, the book resembles a work of mythology, rather than an accurate historical treatise. Consequently, it offers little help for serious thinkers to sort out what really happened during each period of turmoil, so each generation is ultimately left to fend for itself.

The Book of Revelations, referred to in Asimov's tale reminds me of the Upanishads and other Vedic texts that convey to the Hindus knowledge about coming periods of chaos and catastrophe. These writings are seen by them as so sacred that the wisdom of these texts would survive the repeated renewal of the cosmos in endless cycles of destruction and re-creation. After each era of complete obliteration, these works would serve as guides for future civilizations.

It is romantic to imagine the period of collapse of the cosmos as presenting a magnificent suspenseful drama of a universal civilization forced to confront its ultimate doom, as in Asimov's "Nightfall." But, in reality, such a turnabout would have one of two possible effects, neither of which would be directly noticeable.

The first possibility is that the epoch of universal contraction would be simply a natural continuation of the current epoch of expansion, albeit with distant galaxies' light blueshifted instead of redshifted. No immediate local effects would be felt by this transition; there would be no sudden "boom" as the universe's dynamics changed direction. Rather, the Earth would continue merrily in its course, orbiting the sun unhindered as if nothing at all had happened (once again, assuming that the Earth and sun were still in existence).

Because the transition from expansion to contraction would take place over eons, there would be little to mark the changeover. The scale factor of the universe would continue to slow down in its growth, halt its growth for the briefest moment, and then gradually start to decrease. Millions of years later, the effect would be noticed by astronomers when they viewed galactic spectra.

The second possibility for the collapse of the universe might seem on the surface more shocking, but would, in fact, lead to a change that, by its very nature, would never be noticed at all. Some theorists have speculated that time itself would reverse its course during a contracting phase. Following the instant of the scale factor's reversal all events that had occurred before that moment would take place again in backward order. But because human memories would reverse as well, this transformation would never be observed.

The notion of universal time reversal isn't a new idea at all. As far back as Plato's dialogue "The Statesman," there have been theories that time might someday run backward. Plato imagined that time flows forward because of the guiding influence of a god. If this divine will were ever disturbed, all people, animals, and other animate beings would necessarily draw to a stop, reverse tracks, and then grow younger. In other words, in Plato's view, if it weren't for heavenly intervention, the "arrow of time" (the imaginary pointer indicating the course of time's flow) would turn and point in the opposite direction.

The nature of time's flow during the contracting period has been a topic of considerable debate in cosmology in recent years, with some scientists echoing a version of Plato's idea and others stressing the opposite. A number of prominent physicists, most notably the Englishman Thomas Gold, coauthor of the steady-state theory of cosmology, feel that a direct relationship exists between a number of postulated arrows of time, including the thermodynamic (law of entropy) and the cosmological (Hubble expansion) arrows. If one of these arrows changes direction, they believe, all of them must. Other scientists, however, refute this claim and assert that these arrows must be independent, and that no reversal of events would occur during a collapsing phase.

The Currents of Time

To call some physical parameter an "arrow of time," one must conclusively show that it always behaves one way in the future direction and the other way in the past. There are a number of candidates for this designation, and it is unclear how they are all related. Some of these arrows, such as the thermodynamic, are considered more natural, perhaps because they are better understood, while others, such as the so-called quantum mechanical (the direction of collapsing quantum wave functions), have been treated much more speculatively.

The law of increasing entropy, as we've discussed, provides a highly suitable gauge of forward time. If you were to film an event that contained a scene of increasing disorder—china plates shattering on a floor, for instance—and then were to play the film backward it would be obvious that something was amiss. Because shards of china would never assemble themselves into solid dishes on their own, the backward version of the film, presenting such an image, would seem ludicrous. Thus the thermodynamic arrow of time, based on the idea that disorder increases only in the future direction, is quite a visually obvious measure.

Another gauge of temporal directionality, the "arrow of conscious awareness," is much harder to define in precise language. We each have a "gut feeling" that time is moving ahead, a sense that the world is progressing in a unidirectional manner and that we are racing forward headlong into the future. Yet how do we scientifically explain this perception?

Moreover, it is frequently the case that this spell is broken—that time seems no longer flowing forward. Often, in dreams, or other altered states such as hypnosis, sedation, or euphoria, a sense of disconnection from the rhythm of linear cause and effect takes place. Experiencing such a transformed state of awareness, one might develop the sensation that one could jump forward or backward in one's own chronology, even escaping from the currents of time altogether.

Some philosophers, such as Costa de Beauregard, have speculated that the sensation of forward passage in time is a supreme illusion, and that the essence of the universe is four dimensional. In this notion, called the "block universe" model, space–time is a single unchanging entity through which our conscious mind imagines it is traveling. According to this concept, when general relativists divide space–time into spacelike slices, they are simply applying a subjective carving knife to objective reality.

In this view, if one could stand outside of time, it would be possible, in theory, to perceive the entire history of the cosmos at once. Like the protagonist of Jorge Luis Borges's remarkable short story, "El Aleph," one could then gaze into a sort of crystal ball and instantly view everything that is to happen. Nothing would ever be a surprise ever again.

Could it be that in our dreams we can actually travel through time? The English speculative philosopher John William Dunne has suggested this in his controversial treatise, *An Experiment with Time*, published in 1938. The Australian aborigines have long believed that the mind journeys elsewhere during dreams, wandering through an altered chronological reality called dreamtime. But few scientists would concur with these views.

Another view of the origins of the psychological (consciousness) arrow of time maintains that our perception of forward motion stems from the information processing functions of the brain. This is the belief of many physicists, most prominently Stephen Hawking. As the brain consumes data related to the outside world, it continuously converts disordered material (undistinguished gray matter) to ordered states (memory storage). In the process, waste energy is necessarily produced and entropy increased. Therefore, information can only be stored, and memory accrued, in the direction of entropy gain. Our perception of time moving forward is a direct result of the continual increase in the size of our memory banks. Hence, in this view, the psychological arrow of time is derivative of the thermodynamic arrow.

The only difficulty that I have with this concept involves the possibility of memory erasure and its effect on time perception. Often with age, disease, or accidents, memory is lost. Yet no one in the history of medicine has ever reported the sensation of time moving backward. One might even imagine, in theory, rigging up a means in which a human memory is drained in such a manner that the entropy of the rest of universe is slightly decreased. This could be done if the increase in disorder of the brain more than balanced out the decrease in disorder of the outside world. I can't imagine, though, that an unfortunate victim of such a ghastly experiment would experience backward time travel, not even in his or her imagination.

Another postulated measure of time's directionality, besides the thermodynamic and psychological gauges, is the quantum mechanical arrow, which relies on the fact that quantum wave functions tend to collapse in a unidirectional manner. In quantum theory, each particle is represented by a probability distribution called a *wave function*, which depicts the likelihood of the particle being located at any particular place, as well as all other information that can possibly be known about the particle. Until a measurement is taken of the exact location of the particle, it cannot be said to be situated any place in particular; its physical presence, as indicated by the wave function, is spread out over space. The wave function is said to be in an uncollapsed (free) state.

Now imagine that an observer comes along and asks the question, "Where is the particle?" A measurement is then taken of the tiny object's position. Instantly, its wave function is said to collapse into a well-defined particle position state. No longer is the wave function spread out over a number of position values; it is now precisely located at one particular set of coordinates.

According to quantum theory, wave function collapse occurs not just for position measurements, but for all measurements of physical parameters. Collapse is inevitable, instantaneous, and irreversible; it takes place during, but never before, the act of observing any quantity. Therefore, because wave functions always collapse toward the future, the direction of quantum collapse is often seen as a natural arrow of time.

Yet another arrow of time derives from the theory of electromagnetism. When an electron gains or loses energy, it either absorbs or radiates light, with one process being the exact time-reversed version of the other. Theoretically, spontaneous absorption should take place with equal likelihood as spontaneous emission. In practice, however, a free electron left to itself radiates, but never absorbs, light energy. Light propagates forward, not backward, in time. Thus the tendency of radiation to be emitted provides a natural arrow of forward time.

There are other proposed arrows of time, including evidence for time asymmetry in kaon (a type of boson) decay, a discrepancy discovered in the 1960s by American physicists J. W. Cronin and Val Fitch. It is unclear how this subatomic arrow would be related to the other, large-scale

Reverse Performance

arrows. Finally, there is the most controversial arrow of all, the directionality of the Hubble expansion of the universe. Here we walk on shaky ground. Either the Hubble expansion is the key to everything or it is an indication of nothing related to time's arrow at all.

Clearly, at present, the expansion of the universe proceeds forward in time; galactic spectral redshifts, not blueshifts, occur as the cosmos undergoes its future-directed evolution. In other words, if one were to film the history of the universe, and then play it reversed, it would be obvious, from the behavior of galaxies, that the movie was running backward in time.

But is there a profound significance to this time asymmetry, or is it just coincidence? There are many natural phenomena that may evolve in one direction for a long period and then suddenly reverse course; the behavior of the weather comes to mind. Often it seems that winters keep getting colder and colder, for a few years, and then suddenly there is a mild one. Because of the ever-present possibility of reversal, no one would argue that the weather provides a natural arrow of time. Perhaps the universe might either expand or contract in the time-forward direction, with no connection at all between the temporal and scale factor behaviors? We just happen to be in the expansion phase; we might as well be in the contraction stage, according to this view.

Thus, we are presented with two options. If the cosmological (Hubble expansion), thermodynamic, and psychological arrows of time are all directly connected, then it is no coincidence that time moves forward during the expansion phase. The alternative is to imagine no direct link between these arrows, in which case one must explain the forward direction of time during expansion through different means.

For now, let's assume the former and explore what might happen if the direction of expansion or contraction were tied to the direction of time's other arrows.

Playing It Backward

If the thermodynamic, psychological, and cosmological arrows of time were indeed directly linked, then we would have no choice but to assume that the entropy of the world would decrease and that time would proceed backward during a universal collapse. This transformation would occur without warning; the direction of time would immediately reverse itself as the scale factor of the universe began to decrease.

The initial moment of contraction would represent the extinction of human history, and yet it would go completely unrecognized. Our thoughts, feelings, and actions would simply operate in the backward direction, erasing all that has already taken place. Like actors on a videotape placed in a VCR and set for reverse play, we would reenact our lives over again backward, unaware that anything was amiss. From the first instant of contraction, all of human history would devour itself, ravenously consuming all scraps of memory, leaving not the merest morsel of proof behind that humankind ever existed.

Life would be quite peculiar in the contracting phase; yet the joke would be lost on us. Take, for example, breakfast, which would become the last meal of the day. We would arrive at the breakfast table fully sated, with no appetite at all. While the hands of the clock turned counterclockwise, we would sit emptying our mouths, unreading our newspapers, and watching our coffee cups fill up with java. Pretty soon our coffee would be too hot to drink. Meanwhile, on a sizzling pan nearby, eggs would unscramble themselves, ultimately returning to their shells. Rising up and walking backward to a cutting board, we would turn around, untoast bread, and then unslice it into a loaf. Finally, we would leave the kitchen, now with a hearty appetite.

There have been a number of science fiction tales that have dealt with the situation of worldwide or individual time reversal. These fascinating stories provide us with further indications as to what a collapsing, time-reversed universe might look like to an outsider.

In a well-known novel by Philip K. Dick entitled *Counter-Clock World*, published in 1967, the Earth, in 1998, is plunged into a reverse time period, called the Hobart phase, which causes the dead to rise up from their burial places, older people to become younger, and the terminally ill to become healthy once more. While on this world disgorging replaces eating as the most common thing to do at the dinner table, on now-inhabited Mars nothing is affected; all processes take place as normal. Unlike the possible effects of a shrinking cosmos, the bizarre manifestations of the Hobart phase turn out to be purely terrestrial phenomena.

In *Ubik*, another novel of this sort by Dick, a mysterious accident causes all things to begin reverting to their earlier forms. All modern instruments of technology such as cars, refrigerators, and other appliances evolve backward to earlier models (Studebakers, Model T's, early Frigidaires, etc.) It turns out, in the end, that only a specially made spray called Ubik can stop this reversal of forms. This is not, strictly speaking, a time reversal story, but it follows similar lines.

Might an individual's own timeline run backward in front of all others to see? In a rather glum tale by Ian Watson, *The Very Slow Time Machine*, this theme is vividly explored. A capsule is discovered, appearing out of nowhere, at the National Physical Laboratory. Inside this windowed device, which can be only opened from within, is a man whose actions are completely time-reversed. At first, he appears old and clearly quite insane. But as time progresses, he becomes younger and seemingly sane. He holds up numerous signs to communicate, requesting to be left alone in the capsule.

The discovery of such a machine stimulates a concerted effort to understand the mechanisms of time reversal. Finally, after much research, it is found that in order to journey forward in time, one must first experience a long period of backward time travel to propel one ahead in a sort of slingshot effect. That is precisely what the hapless fellow in the box has been doing—and, as witnessed by the first view of him, he has gone stark raving mad in the process.

In Roger Zelazny's short story "Divine Madness," another unfortunate figure experiences the lunatic effects of time reversal, in this case through a series of strange seizures that propel him backward in time for brief periods. After a number of these painful episodes, in which he experiences undrinking cups of coffee, unshowering and walking backward numerous times, he decides to commit suicide. But, alas, as soon as he is dead and buried, his corpse arises from the grave, reforms into a living human being, and begins the process of backward motion all over again. He has become the living embodiment of the Sisyphus myth, reliving his actions over and over again.

Finally, a one-page short story by Frederic Brown, entitled "Experiment," sums up this entire topsy-turvy genre in a few paragraphs. An inventor, Professor Jones, develops and tests a machine that can reverse time. In a humorous plot twist typical of Brown, as soon as Jones presses the button operating the time-reversal machine, the story's nar-

Reverse Performance

ration starts to reverse itself: "Pushing a button as he spoke, he said 'This should make time run backward run time make should this' said he, spoke he as button a pushing."[1]

Thus, ironically, Professor Jones, through his arduous efforts, has only succeeded in erasing himself from history. His painstakingly constructed invention has removed all traces of its own creation. It is truly an ourobouric device: a machine that has devoured itself.

It is interesting to note that in virtually all of these stories, there is a contrast made between the bizarre actions of those unlucky enough to be traveling back in time, and the normal life patterns of all others, including the reader. In reality, though, if the universe's temporal arrow were to reverse suddenly, there would be no uninvolved witnesses present to be either horrified or amused. While unshowering, driving in reverse, or unslicing a loaf of bread, all thoughts would proceed backward as well, so nobody would even know the difference between the strange events taking place and normal life. As Paul Davies, of the University of Newcastle-upon-Tyne, puts this:

> A human being in a reversed-time world would also have a reversed brain, reversed senses and presumably a reversed mind. He would remember the future and predict the past, though his language would not convey the same meaning of his words as it does to us. In all respects his world would appear to him the same as ours does to us.[2]

The microscopic world, on a quantum level, would presumably be affected by universal time reversal as well. According to Nobel laureate Murray Gell-Mann of the California Institute of Technology and James Hartle of the University of California at Santa Barbara, the theory of quantum wave collapse would be made complete by assuming that time runs forward during expansion and backward during collapse. They have devised a "time-neutral" quantum theory, which exhibits the same behavior in both temporal directions.

In Gell-Mann and Hartle's system, the initial and final states of wave functions are incorporated into a single time-symmetric mathematical approach, which shows how the initial might evolve into the final, as well as how the final might evolve into the initial. Both processes are assigned equal value, with the former taking place during the opening half of the universe's life and the latter during the contracting half. This theory, if valid, would remove the seemingly built-in time-asymmetry from quantum wave collapse. One must simply assume that the law of cause and effect would reverse itself during the twilight years of the cosmos.

Gell-Mann and Hartle hope that their time-neutral theory will remove some of the "philosophical baggage" that they believe is associated with the standard model of wave function collapse, namely, a preferred time direction. As Hartle puts this:

> It is not an arrow of time which is fundamental, but rather the fact that the universe is simply at one end and not at the other.[3]

In other words, chaos will revert to order in the final stages of the cosmos. Gell-Mann and Hartle contend that quantum collapse and the arrow of time must be considered relative to whatever is the current state of the universe. It is only because we are presently expanding, and entropy is increasing, that quantum wave functions are collapsing.

Time Without Boundary

For a long time, Stephen Hawking was a strong advocate of the notion that the collapse of the cosmos would bring about a reversal of the hands of the clock. But after consulting with physicists Don Page and Raymond Laflamme, carrying out a number of tedious calculations, and doing a great deal of soul searching, he performed a turnabout himself, and became a supporter of the idea that time during universal shrinkage, rather than switching modality, would just continue on its merry course. Hawking speaks of this change of heart in *A Brief History of Time*:

> The beginning and end of time can be very different from each other. I was misled by work I had done on a simple model of the universe in which the collapsing phase looked like the time reverse of the expanding phase. However a colleague of mine, Don Page, of Penn State University, pointed out that the no boundary condition did not require the contracting phase necessarily to be the time reverse of the expanding phase. Further, one of my students, Raymond Laflamme, found that in a slightly more complicated model, the collapse of the universe was different than the expansion. I realized that I had made a mistake: the no boundary condition implied that disorder would in fact continue to increase during the contraction.[4]

Hawking was not the first to suggest that the shrinking of the cosmos might be different from a time-reversed picture of its expansion. At least a decade before Hawking's turnabout, Roger Penrose of Oxford University, a former collaborator, proposed his own cosmological explanation of the law of increasing entropy, involving very disparate scenarios for the beginning and end of the universe.

In Penrose's view, the thermodynamic and other arrows of time stem from a strong discrepancy in the boundary conditions of the cosmos, that is, the initial and final states. While the Big Bang was, according to the Oxford mathematician, a highly regular and ordered event representing the condition of zero entropy, the Big Crunch demise of the cosmos will be an extremely disordered occurrence, embodying the state of maximum entropy. This built-in cosmological dissimilarity, derived from a structural component of general relativity, is what drives the events of the universe to proceed forward in time, not backward.

Hawking's model employs universal boundary conditions as well to explain the nature of time's flow during the later stages of the cosmos, though, ironically, his boundary conditions are called the "no-boundary condition." The no-boundary proposal, first suggested by Hawking and Hartle, and later modified by Hawking, Page, Laflamme, and Lyons, removes all uncertainty about the nature of the initial and final states of the cosmos by boldly assuming that the timeline of the universe is contained, but without limit, like a circle.

The no-boundary condition was developed with a profoundly atheistic spirit. It removes all reference in cosmology to a divine creator by assuming that the universe never began. Thus, there is no need to imagine a deity selecting the initial conditions of the cosmos, because there was no beginning state.

One might wonder why Hawking is so concerned about the image of God choosing initial parameters for the universe. Hawking has always felt that one of the flaws of the standard cosmological model is that it doesn't explain what happens before the moment of creation. Moreover, since there are innumerable possibilities for the initial state of the universe, the Big Bang model forces us to assume that one of these choices was favored over the others. For example, one possible model of the opening moments of the universe is an open Friedmann cosmology; another is a closed anisotropic cosmology. Only one of these possibilities represents our own world; somehow, the rest were rejected.

Inflation would help a bit with this dilemma. The inflationary era would obliterate all traces of the peculiarities—

anisotropies and inhomogeneities—of the initial state of the universe. It would act on spatial irregularities like an immense washer–dryer full of the strongest possible detergent, smoothing out all wrinkles, bleaching out all stains. We would be left, after an inflationary era, with a fully laundered universe, substantially free of blemishes, as the COBE findings do indeed indicate.

Yet, even with inflation, we would still face the question: Who or what chose the original outfit for the cosmos, wrinkles and all? Was it the roll of the dice, God, neither, or both? (Einstein, however, has suggested that God does not play dice.)

This issue is exacerbated by the likelihood of an initial cosmic singularity. A singularity is a point that mathematically represents the edge of coordinate space. Hence, it denotes the very edge of knowledge, the point beyond which no information can be obtained. Hawking and Penrose, in a famous theorem, have proven that any classical (nonquantum) solution to Einstein's equations modeling an expanding universe must possess an initial singularity. Thus there appears, in standard cosmology, to be an infinite number of possible scenarios for the first moments of creation, all depending on what comes out of the singularity.

Physicists deplore the presence of singularities in models, because they leave too much free reign in the choice of solutions. It has been shown that black holes must possess singularities as well, fortunately hidden behind cloaks of darkness. Unlike the Big Bang's early state, with its *initial* singularity, black holes encompass *final* singularities, which are, in many ways, philosophically more palatable. It is easier to imagine organized matter being drawn into oblivion, rather than mass being produced from oblivion. Perhaps in our rather jaded society, we can more readily conceive of unlimited destruction than contemplate unlimited creation.

Creation implies a creator, and that is why the presence of an initial singularity is so evocative of the Judeo-Christian-Islamic image of Genesis. "Let God choose a universe most pleasing to Him and let it explode in a Big Bang" is the hybrid part-religious, part-scientific model that is often declared. This is pleasing to theologians and comfortable for physicists.

Nevertheless, Hawking and his colleagues have now proclaimed, "Let there be no true beginning." In effect, what they have done is to remove any pretext for God to play a role in physics. Hawking has even been so bold as to discuss these ideas during a conference at the Vatican.

The no-boundary condition manages to avoid the mind-boggling singularity that caps the standard Big Bang model. Basically, to remove this unsavory element, a bit of mathematical manipulation is necessary. The time parameter in the general relativistic metric that characterizes the universe is replaced with a so-called imaginary time coordinate. This transforms all four-dimensional metric distances so that they now are qualitatively similar to ordinary distances; namely, in the new coordinates, metric distances are always positive. This transformation accomplishes the task of removing the singularity.

Let us explore the difference between real time and imaginary time. In real time, according to an earthbound observer, if you managed to travel at the speed of light (say, you arranged somehow to turn yourself into a photon) and cover a distance of 186,000 miles, your trip has lasted one second. In imaginary time, on the other hand, your trip has taken the square root of minus one second.

Reverse Performance

Clearly, we live in real time. We never observe our stopwatches to record the durations of sporting events in terms of imaginary numbers. The use of imaginary time in the no boundary model isn't a true measure of the duration of cosmic events, but rather just a convenient way to avoid singularities. The prescription for this is first to express the evolution of the universe in terms of imaginary time, then to transform the resulting equations back into real time. The history of the Big Bang can then be mapped out as a function of real time.

Hawking and his colleagues have made the explicit assumption that the universe is closed and bounded (finite). Following Friedmann's ideas, a positive spatial curvature is utilized. In this approach, an inflationary epoch, with its consequent flattening of space, can be readily accommodated by assuming that the universe at present has a value of omega that is slightly greater than one, just enough for closure.

One might wonder what would happen if one were to apply the no-boundary condition to a space–time of negative curvature. It turns out that one must use a positive curvature space–time with the no-boundary proposal, or else poor mathematical results are obtained, leading to a clear incongruity. G. Oliveira-Neto, of the University of Newcastle-upon-Tyne, England, in a recent calculation, has shown that negative curvature would lead to an initial singularity, directly contradicting the spirit of the no-boundary assumption. So, it seems that the closed Friedmann universe, exhibiting positive curvature, is the most suitable candidate for Hawking's scenario.

The no-boundary model posits a universal evolution that is reminiscent of the behavior of the lines of latitude on Earth as one travels away from the North Pole, down to the equator, and then farther down to the South Pole. Consider the circumferences (lengths around the globe) of each of these lines to represent, in succession, the scale factor of space for consecutive times. Here, the North Pole signifies the initial moment of the Big Bang, when the universe has zero extent. As one goes south, these global rings get wider and wider—corresponding to an expanding cosmos—until one reaches the equator. This signals the moment of maximum expansion, the turning point of the scale factor. Traveling farther south, the circles now grow smaller, representing the recontraction of space. Finally, one reaches the South Pole: the Big Crunch point.

We can see now how the initial singularity is avoided. There is a perfectly unambiguous and continuous way of traveling right through the North Pole and ending up on the other side. Hence, the North Pole, representing the dawn of time, cannot be considered a singularity in this model.

Originally, in proposing the no-boundary notion, Hartle and Hawking supposed that the similarity of the North and South Poles meant a complete symmetry for the model in time. There seemed to be no qualitative difference between these polar extremes, so they supposed that the approach to the Big Crunch would simply be a time-reversed version of the approach to the Big Bang. Hence, they initially concluded that the chronicle of events would roll backward during the latter half of cosmic history.

However, Hawking, convinced by Page and later Laflamme, soon realized that imaginary time is not the same as real time. The global lines of latitude correspond to degradations of imaginary, not real, time, and one must distinguish between the two. In particular, even if *imaginary* time were symmetric, as the global model suggests, it is not necessarily the case that *real* time would be symmetric as well. Thus, they concluded, there is no *a priori* reason to assume that time would reverse itself in the latter half of the no-boundary universe.

Laflamme's recent calculations at the Division of Theoretical Astrophysics at Los Alamos National Laboratory tend to support, in fact, the view that real time would be unidirectional throughout the entire history of the cosmos. This conclusion is based upon a detailed study of the growth of inhomogeneities in the universe, subject to the no-boundary stipulation. Any minuscule irregularities present in the first stages of the cosmos, he found, would continue to accrue, even if the cosmological expansion were to reverse itself. Consequently, entropy would continue to become greater throughout all universal eras, from the Big Bang to the Big Crunch. Because the arrow of thermodynamic time and, as Laflamme and Hawking, among others, believe, the arrow of psychological time are linked up with the increase of entropy, these arrows would point in the same direction forever.

One final puzzle is why, if the universe possesses both expanding and contracting phases, did intelligent life begin in the expanding stage and move forward in time, rather than begin in the contracting stage and move backward? The answer, as Hawking relates in his book, is that intelligent life needs a great deal of order, and thus a low entropy state, to thrive. Hence, it is conceivable that life would begin in the early expanding years of the universe, with its abundance of usable energy, but inconceivable that life would form in the twilight era.

So, here, in the proposal of Hawking, Laflamme, and company, we are presented with the least traumatic model of the behavior of a Friedmann closed universe. There would be no sudden catastrophes, save the end, and no temporal reversals. Apparently, from the instant of the Big Bang, through the stages of growth and collapse, until the Big Crunch itself, nothing would alert us to the fact that the cosmos was developing in this manner. Clocks would never run backwards, eggs would never unscramble themselves, and presumably there wouldn't be riots—or antiriots—to herald the falling of the sky. And when the Big Crunch finally arrived, the universe would simply consume itself, quickly and painlessly.

Generally speaking, astronomers haven't expressed much enthusiasm for the no-boundary model. Perhaps they find it to be too speculative for their tastes, with its far-reaching conclusions almost impossible to test. Nevertheless, with its bold premise that the universe is completely self-contained in space and time, it is certainly a fascinating cosmological approach to ponder.

The World That Devoured Itself

Over six millennia ago Chinese artisans were etching onto ornate vases the symbol of the self-devouring serpent *ourobouros*. Centuries later, the Elamites of Susa (Iran) and the Jainas of India were producing similar versions of this pattern in their artistic imagery. And *ourobouros* motifs are prevalent in numerous relics from ancient Egypt.

The no-boundary model of the universe bears amazing resemblance to cosmological notions suggested by these artifacts. Somehow these early craftsman managed to anticipate by many centuries the idea of space begetting itself. Without a doubt, the time-worn image of *ourobouros* is an incredibly accurate symbol of Hawking's modern concept of self-generated universal creation and destruction.

Reverse Performance 245

One might think that this remarkable resemblance is more than a coincidence. Did ancient astronauts or time travelers from the future bear Hawking's message of a world without boundaries back to these early artisans? Might the *ourobouros* symbol be proof that our ancestors possessed extraordinary knowledge of the heavens?

Absolutely not. There are essentially only a finite number of ways (give or take small variations) of imagining the creation of the universe. So rough parallelism of ideas is not at all surprising.

One of these ways of viewing time is to picture an omnipotent God saying "Let there be light" sometime in the past. This case corresponds scientifically to a Big Bang model in which the universe was created in time. One might point to some particular moment and say, "This is when natural history began. The rest is a mystery." These world views share the fact that they are linear theories of our origins.

Another way of looking at the cosmos is to picture it first generating itself and later consuming itself. There is no space beyond the space we can picture and no time beyond an all-encompassing but finite time. Philosophically, this is quite attractive, since it makes no assumptions about any time intervals beyond creation or Armageddon. There simply is no moment before the beginning or after the end. This happens to be both the *ourobouros* motif and the Hawking model.

Finally, there is the idea of an infinite succession of cycles, a notion present in both Hindu texts and in the modern theory of an oscillating universe (repeated Big Bangs, each followed by a Big Crunch). Once again, there is parallelism, but only because there is basically a finite number of ways of looking at time: linear, cyclical, self-creative, etc. These religious and scientific ideas are coincidentally the same.

With that all said, let me emphasize the enormous evocative power of these religious narratives in describing our world. Science can describe our cosmos mathematically, but somehow its accounts lack lifeblood. Religion is flowing with life, though it cannot provide the stone-cold equations of proof. To develop a living, breathing metaphor is a beautiful thing, and we need all the beauty we can get.

In linear theories of the cosmos, a creator or creative force outside of time is a necessary element. For instance, in the standard Big Bang model, it is essential for completeness to imagine some way of the universe coming into being. One might imagine the biblical God, existing before the world began, accomplishing this task.

By contrast, there is no room for a pre-creation God in the no-boundary or oscillating models of the universe. In the first case, the cosmos created itself, and in the second, the universe was reincarnated from an earlier version. Therefore, may I suggest as the appropriate metaphor for these models the ancient image of a God outside of mundane time and outside of creation, existing in what the great Romanian religious scholar Mircea Eliade (of the University of Paris at the Sorbonne and the University of Chicago) called "sacred time."

In this imagery, God (or the gods) stands far removed from the ordinary dimensions of space and time. The universe came into being, not at some particular moment in its history, but all at once, outside of time, with its physical laws determined by its Designer. And (in this metaphor) like *ourobouros*, it is programmed to regenerate itself continually.

AFTER THE CRUNCH

> *From a moral standpoint the conception of a cyclic universe, continually running down and continually rejuvenating itself, seems to me wholly retrograde. Must Sisyphus forever roll his stone up the hill only for it to roll down again every time it approaches the top? That was a description of Hell. If we have any conception of progress as a whole reaching deeper than the physical symbols of the external world, the way must, it would seem, lie in escape from the Wheel of things. It is curious that the doctrine of the running-down of the physical universe is so often looked upon as pessimistic and contrary to the aspirations of religion. Since when has the teaching that "Heaven and Earth shall pass away" become ecclesiastically unorthodox? Ah, who will count the universes that have passed away or the creations that have risen afresh, again and again, from the formless abyss of the vast waters?*
> —**Arthur Eddington, 1933 Cornell Lecture**

The Rhythm of Eternity

When I was a child I remember standing on the beach in Atlantic City just before dawn, watching the perpetual motion of the dark ocean waters. In the eerie silence as the sun broke through the horizon, I could hear the pounding rhythm of the surf, the constant rising, churning, and pulling back of the sea. The dome of heaven had become one vast conch shell against my ear, amplifying the incessant crashing of the waves. I stood drawn in and hypnotized by the ceaseless pattern of motion, the interminable repetition of the vast cold froth.

With our telescopes we can detect the galaxies receding, the redshifted light as a telltale sign of the cosmic undertow. In this process we know that the dark, star-speckled sea above us is becoming vaster and colder, and more and more hopeless for intrepid beams of light to span. We send out our *Voyager* messages in a bottle, but still we hear no answer; we know so little of our island galaxy, let alone the gravitational tides that span all of space.

Might the tremendous force of gravity send all of vast space flooding in and gushing out repeatedly like the waves lapping a jetty? If so, would there ever be an end to this series of universal implosions and reexplosions? The answers are stored in the very fabric of the cosmos, bound up in a tangled set of equations. When the truth is finally known, we shall be able to peer across the waters of cyclic eternity, having solved, at last, the riddles posed by the ancients.

Beyond the Omega Point

When Friedmann proposed his cosmological models, he provided no information about what might happen before the first moments of what we now call the Big Bang or after the final death throes of what we now call the Big Crunch. Yet it is only natural to ponder what might have preceded the "initial" explosion and what might succeed the "ultimate" collapse.

Hawking's no-boundary condition is one distinct possibility. The universe might have simply generated itself in a flash of autocreativity, and be destined to engulf itself someday. But because this somewhat controversial model is untested, we must certainly consider the full realm of possibilities, including more traditional explanations.

In the case of a closed universe, it is compelling to imagine that the explosion that set off the Hubble expansion in our era (or set off the seeds of inflation, as the case may be) was triggered by the contraction of the universe during a previous era. In other words, another period of the cosmos, with perhaps a similar development of stars, galaxies, and even life-forms, might have preceded our own. And maybe, in that case, another cosmic era preceded that one, and so on.

Extending this logic, let's imagine what succeeded the instant of complete contraction. Some science writers, most notably John Gribbin, have taken to calling this seemingly final moment, the *omega point*. The name is taken from a strange mixture of religious and scientific theory proposed in the early part of this century by the eccentric French-born Jesuit theologian Pierre Teilhard de Chardin.

Teilhard optimistically prophesied that the end of the universe, which he loosely equated with the Second Coming, will signify the culmination of the creative powers of humankind. Evolution, he supposed, will cause the human race to develop rapidly in complexity, acquiring more and more knowledge. Ultimately, according to Teilhard, our peoples will spread out through the cosmos and acquire the skills to control the entire universe. The result will be a unity of humankind and nature that leads to the contraction of space into one great ball of energy, called the "noosphere."

This noosphere further contracts, leading to the omega point, named for the final Greek letter.

Gribbin points out, in his book *The Omega Point*, the appropriateness of this phrase in describing the moment of complete collapse:

> 'Big Crunch' is an ugly term which hardly seems appropriate for so important an event as the end of the universe. But there is no convention as yet for a label of the moment of destruction at the end of time, and I am free to borrow the term 'omega point.'[1]

If a succession of universal epochs cyclically anticipated our own Big Bang, then one might imagine myriad cosmic ages following the omega point to come. Thus, the *scientific* omega point represents both an end and a beginning, an end of one cycle and beginning of the next. This notion of an infinite series of Friedmann episodes of expansion, contraction, followed by reexpansion and recontraction is called the *oscillating universe theory*.

Considered from the standpoint of energy conservation, the oscillating universe model seems quite intuitive. It is natural to imagine energy being siphoned from one spatial epoch to the next, the contraction of the former driving the repulsion of the latter. No energy is gained or lost; it is just converted from one form to another.

One poignant issue that arises with this approach is the survivability of the human race (or anything material for that matter) during the omega point. The possibility of survival seems far more reasonable if one imagines contraction to a finite region, instead of complete collapse to a mathematical point. The image here is one of an elastic and durable balloon being squeezed into a tiny ball, and then springing back again to full size.

Perhaps, according to this bouncing universe approach, at the last moment before oblivion a cohesive force

After the Crunch

would stop the universe from fully contracting to a point, and cause it to "bounce" back again to its maximum extent. The Hubble expansion characterizing our present epoch would represent, then, the rebound effect of an earlier episode of this sort.

Clearly, the bounce-type universe offers more of a chance for the human race to endure. We would merely have to come up with a way of surviving the harsh, dense conditions of a collapsed cosmos, rather than planning how to weather a pointlike existence. Real estate prices might go up, as space would become a premium for a time, but shrewd investors could count on a new era of expansion to come.

There's one possible glitch, however. It was thought until recently that bouncing of this sort would be precluded by the Hawking–Penrose singularity theorems. Hawking and Penrose found that it would be literally impossible, within the classical framework of general relativity, for a Big Bang to arise from, or a Big Crunch to withdraw into, anything other than a point. Any scenario of collapse and reexpansion, they emphasized, must include the crushing uniformity of a point singularity.

As discussed, the no-boundary approach was designed to avoid this problem. But even in the more conventional oscillating universe model, there is considerable hope for reprieve. Hawking and Penrose's results apply only to *classical* gravity; a complete picture of the omega point must include quantum effects. In any high-density situation—the end of the cosmic era being no exception—we must reckon with quantum gravitational corrections to the relativistic classical solutions. It is conceivable that these effects would provide just the bounce needed to avoid singularity. More needs to be known about quantum gravity before any conclusion can be drawn.

The Entropy Crisis

Probably the most detailed analysis of an oscillating universe scenario was performed in 1934 by the American physicist Richard Tolman. Tolman was particularly interested in the thermodynamic implications of cosmic cycles. Unlike Thomas Gold, he assumed that time would move forward and entropy would accrue during the repeated periods of contraction. He avoided consideration of singularities by presuming that physically realizable universes (unlike what he supposed to be overly simplified models) would contain no such mathematical peculiarities.

Tolman's conclusions turn out to be most disturbing. Yes, there can be spatial oscillations, he found, but there is a keen cost to be paid. The entropy built up from one cycle must be carried over into the next. Because this means that each Big Bang must begin in a more disordered state than the previous one, clear limits are placed on the physical parameters of successive eras.

For one thing, as Tolman calculated, each cycle must be of longer duration than its predecessor. This is due to the accumulation of starlight from era to era, causing additional radiation density. Eventually, Tolman supposed, space would be full of this excess disordered energy.

Also, in the beginning of each cycle, the expansion of the universe would start out faster than at the start of the previous cycle. This would mean a greater maximum value of the scale factor, each time, at the turning point before universal contraction. Consequently, the collapse would take place, in successive eras, more and more forcefully.

After the Crunch

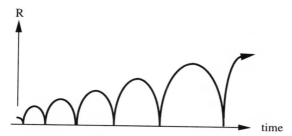

Figure 23. Tolman's oscillating universe model. The scale of the universe is plotted as a function of its age. Note that each "bounce" of the universe lasts longer and achieves a greater amount of maximum expansion than its predecessor.

The end result of Tolman's calculations is sobering for those who harbor the immortalists' hope that the universe passes through an infinite series of oscillations, without beginning or end. Based on the current conditions of space, particularly its background temperature of 2.73 degrees Kelvin, one can estimate the number of cycles that preceded ours to be no more than a hundred.

The picture that emerges from these results is a hybrid of the single Big Bang and oscillating universe models. Sometime, billions of years in the past, the cosmos emerged from nothing in the form of a low-entropy fireball (albeit of much lower temperature than our own Big Bang). Space expanded out from this point and immediately began to collapse, building up entropy in the process. It then expanded over a slightly longer period only to contract once more with slightly greater force. This went on for scores of cycles until the epochs of expansion lasted long enough for stars to form, and finally for intelligent life to form. Eventually, in cycles following ours, the entropy of the cosmos will be too great for life to develop at all, upon which there will be the strange situation of a universe expanding with no intelligent inhabitants to observe it.

One might wonder then how the first tiny fireball originated in this scenario. One intriguing possibility, suggested by Ed Tryon in the early 1970s, is that the universe arose from nothing by means of a quantum fluctuation. In quantum theory, objects often appear from the vacuum spontaneously, and decay almost immediately. This happens because of the dictate, called the Heisenberg uncertainty principle, that mass and energy conservation can be violated for a brief interval; the greater the discrepancy, the shorter the time of violation. Therefore, Tryon supposed, one might picture the Big Bang (more specifically, the first Big Bang) to be a massive fluctuation of this sort.

The Tolman model traditionally has had few adherents. Philosophically, it alienates two large groups: those who theorize an infinite series of cycles and those who believe in a single Big Bang. The former group might wonder: Why a beginning at all? The latter ponders: Why not a single beginning? So, there are few who are entirely pleased by this hybrid.

Moreover, it has been brought to attention by Penrose that there are gaps in Tolman's estimation of the amount of entropy accumulated during cycles. Tolman, working before the analysis of black hole dynamics was developed, naturally didn't include the effect on the fate of the universe of these stellar relics' total entropy. Once black hole entropy is added to the equation (a black hole's entropy is proportional to its surface area), then there seems to be a far greater accumulation of disorder—enough to rule out more than one cosmic cycle. This situation, if it weren't alleviated by other means, would certainly represent a crisis for the oscillating universe theory.

Reprocessing Space

There have been a number of important attempts in recent years to solve the entropy crisis for the oscillating universe. The payoff for a successful solution of this issue would be the scientific possibility of realizing the ancients' dream of eternal cosmic renewal. Otherwise, all hope for a cyclical cosmos would be lost.

One promising scenario for resolving the problem of excessive accumulation of entropy was proposed by Werner Israel of the University of Edmonton, along with several of his colleagues, including Eric Poisson and A. E. Sikkema. During a Big Crunch phase, they surmised, the universe would literally be packed with black holes, the remnants of millions of aged galaxies and other massive celestial artifacts. These black holes would maintain enormous densities, stocked not just with material remains of stars, but also with tremendous quantities of gravitationally blueshifted light trapped within. This blueshifted radiation, its spectrum altered by the strong gravitational fields near black holes, would be far more energetic than ordinary starlight.

As the universe neared its omega point and grew smaller and smaller, all of its constituent black holes would begin to overlap. These mergers would lead to a greater and greater combination of mass. Finally, approximately an hour before the cosmos would have contracted into a singular state, the sheer mass of the increasingly concentrated black hole behemoth would cause an abrupt implosion to occur. In less than a trillion-trillionth of a second, the entire cosmos would become concentrated in a region far smaller than the size of a proton.

The effect of this sudden compression, according to Israel and his co-workers, would be quite suitable for the purpose of cyclic regeneration. First of all, because of the tremendous shrinkage of the black hole conglomerate, a great deal of entropy would be lost. Enough would be dissipated, in fact, that very little would be left as the omega point was approached. Second, the suddenness of this collapse would cause a sort of rebound effect; shock waves (due to quantum mechanical principles beyond the scope of this discussion) would force the universe to expand again. The singularity at the omega point would never be reached. Consequently, the two primary issues plaguing most oscillatory models—the singularity problem and the entropy crisis—would be avoided here. Because the entropy buildup would be so small between oscillations, this model would allow for a very large (but, alas, finite) number of cosmic cycles.

Another proposal for dealing with these issues, involving the late-term reunification of the forces of nature, was advanced in 1982 by Vahe Petrosian of Stanford. He imagined that sometime in the declining stages of the universe all of the forces—strong, weak, electromagnetic, and gravitational—that separated out in the cosmos' early years would unite once more. The net result would be that new fields would be created. The energies of these fields would drive an inflationary epoch in which the universe would rapidly expand again. In this manner, the singularity would never be reached.

Petrosian deals with the entropy question by assuming that we presently reside in a cycle in which entropy has already accrued from the past. Hence, the implication is that

After the Crunch

only a few cycles might be realized in the history of the universe, because heat death—the state of maximum entropy—would be reached after a limited number of oscillations.

Yet another attempt to circumvent the entropy crisis was advanced by John Wheeler in the 1970s. It involves the complete reprocessing of the fundamental constants of the universe during each of an infinite series of omega points. In his scheme (which he no longer advocates), Wheeler proposes that each oscillation of the cosmos would embody a whole new set of conditions, such as masses, gravitational constants, and the like. All physical parameters would be arbitrarily adjusted from cycle to cycle. In this manner, the entropy at the start of each cycle would be reset to an arbitrary value as well.

So, in the Wheeler reprocessing scheme, the nature of the universe, and therefore the prospects for intelligent life, during each cycle would be as random as a coin toss. We would simply be denizens of one of many possible manifestations of the universe, residing here in this particular combination of elements by sheer chance.

Renowned cosmologist Edward Harrison has expressed the essence of Wheeler's model quite well:

> Perhaps, in the ultimate chaos, there lurks the cosmogenic genie who conjures universes into being, and from the primordial ferment spins new worlds of every possible kind.[2]

Recently, however, John Barrow of the University of Sussex, and Frank Tipler of Tulane University, have pointed out several major difficulties with Wheeler's scenario. First of all, it cannot possibly be tested; there is no experimental way of knowing if the universe would be reprocessed during an omega point. Second, if all physical parameters were altered between successive cycles, then it is reasonable to assume that the spatial geometry would be randomly modified as well. This would create a significant chance that each oscillation for a closed Friedmann model could transform into an open model. And because open models don't cycle, such a transformation would represent the end of the whole process. So the number of cycles would be manifestly finite.

Finally, to complete this survey of methods, I should mention an entirely fantastic way of dealing with the problem of entropy buildup. In Isaac Asimov's short story "The Last Question," computers of the future, of increasingly complex design, are successively asked to analyze the question: Can entropy be reversed? Early computer models are stumped by this question. They reply, "Insufficient data for meaningful response." Meanwhile, while new computers are constructed to try to resolve this dilemma, the entropy of the universe continues to grow. Finally, the ultimate computer is created, and naturally it has an answer to the query. In the manner of Genesis, it says "Let there be light," and remakes the entire universe, sans entropy. Thus the entropy crisis is solved by the "divine" intervention of a world-building, entropy-destroying computer. Alas, if only this were possible.

Only the Names Are Changed

In the oscillating universe theory, as pictured by Israel, Wheeler, and others, universal contraction and reexpansion would cause an entirely new phase of the cosmos, with new stars, planets, nebulae, and other features, to come into being. In each cycle, therefore, a totally unique roster of occurrences would be acted out.

This notion of *inexact* cosmic cycles is quite distinct from the Nietzschean concept of *exact* repetition of all events. Nietzsche imagined eventual revival of all persons, places, and things, reenacting the dramas of life as if on a late-night TV rerun.

Then would exact recurrence necessarily be excluded from an oscillatory model? Or might eternal return of all earthly events occur in some future cycle of creation? In other words, is there hope for the sort of "immortality of repetition" envisioned by the great German mystic a hundred years ago.

Recall that there are reasons to believe that under certain circumstances—in particular, a closed system with a finite number of states—duplication of all events would eventually take place. Poincaré demonstrated in his famous recurrence theorem that any bounded set of atoms must return infinitesimally close again and again to any given initial configuration. And, because it is certainly true that a closed universe represents a bounded system, there is reason, albeit small, to suspect that Poincaré recurrence would occur in an oscillating cosmos.

Granted, we are talking about completely different time scales here. A Friedmann-type complete regeneration of the cosmos, including expansion and contraction phases, would take place during a span of tens of billions of years. Raise this figure to its own power, and the lifetime, in years, of a Poincaré recurrence of the whole universe would still not be approached! Countless oscillations would need to occur before the exact conditions of present-day Earth would be replicated.

Still, in light of the eventual certain extinction that is the lot of us mortals, it is gratifying for some to contemplate at least the *possibility* of recurrence, even if it would take nigh eternity to come about. For those who shun religious explanations and believe in nothing but the material realm, this

Figure 24. Frank Tipler, b. 1947. (Courtesy of Frank Tipler and Paul Daigrepont Studios.)

sort of return would provide a conceivable, albeit time consuming, method of "reincarnation."

Unfortunately, for those hoping to be "recycled" back to Earth in an eternal return, a calculation performed by Tipler has lent, perhaps, a death blow to the whole idea. Tipler has shown that Poincaré cycles cannot occur in an oscillating cosmos—at least one that passes repeatedly through point singularities. Using the fact that gravitational forces are always attractive, he has proven what he calls the "no-return theorem," a statement of the impossibility of any two states of a closed universe ever being identical or even arbitrarily close. Basically, in his proof, the infinite curvatures of omega point singularities cause enough divergence of paths to insure that exact repetition of circumstances would be impossible.

For the no-return theorem to work, the caveat here, is that spatial singularities must be present. Tipler has remarked that if a way could be found such that these singular points would never be encountered (that is, a "bounce scenario"), this set of arguments would be voided and Poincaré recurrence would instead be possible. Quantum gravity (a theory of gravitation on a microscopic scale) might someday indicate a clear way of avoiding these singularities, opening up the whole question of eternal return once more. Theorists suspect that in a quantum theory of gravity, tiny random fluctuations of the fabric of space would serve to smooth out singularities, enabling a "big bounce" to take place. In that case, eternal return would at least be conceivable (though extraordinarily time consuming).

Surviving the Crunch

If there is little to no chance for survival of the human race through eternal recurrence of events, perhaps humans might outlive cosmic collapse through other means. Clearly, if the universe were to pass through a singularity, there would be no hope for our species to escape oblivion (assuming that humankind still existed up until the omega point). A bouncing model, which avoided the state of infinite curvature, would offer at least a minimal chance for intelligent life to continue, because our race might in theory develop the means to survive the hostile conditions (intense radiation, strong gravitational fields) of a greatly shrunken cosmos.

Suppose, though, that scientists of the far future determined that it would be absolutely impossible for humankind to withstand the Big Crunch. In that case, alternative methods of escape from our shrinking sector of space would need to be devised. Somehow, we would have to leave our collapsing universe and tunnel into another. This is assuming, of course, that parallel (alternative) universes exist outside our own—a highly speculative notion indeed.

Assuming, for the sake of argument, that there are other universes besides ours, one possible scenario for escape would involve the use of postulated special segments of warped space–time, called *traversable wormholes*. Traversable wormholes are special solutions of Einstein's equation derived in 1988 by Kip Thorne and Michael Morris of Caltech. They are designed to serve as navigable links between otherwise distant or disconnected parts of space, including, perhaps, separate universes.

In a highly speculative scheme for our species' rescue, it has been thought that the human race could travel through one of these spatial shortcuts into another, preferably nonshrinking, part of space. Theoretically, at least, according to Morris and Thorne, wormholes would provide safe and speedy passage from one region of space to another. In practice, however, because of their highly specialized properties, finding or creating one of these hypothetical objects would be extremely difficult, to say the least.

Might the human race, or other forms of intelligent life, use wormholes or other escape routes to endure cycle after cycle of creation and destruction? Perhaps the period of Hubble expansion in which we find ourselves today signifies only the start of a multicycle chronicle of human history?

After the Crunch

The idea that humankind might survive numerous terrestrial and universal cycles is the theme of Olaf Stapledon's classic science fiction epic, *The Star Maker*, published in 1937. In this highly imaginative chronicle of the future, an Englishman learns to travel through time, and witnesses the entire history of the universe, from its Big Bang origins to its Big Crunch demise. Seeking an understanding of what will happen following the death of the last star, he encounters the godlike spirit controlling all of cosmic evolution, the Star Maker.

The Star Maker, physically located at the center of the closed, curved hypersphere of space, has the supreme power to expand or contract the cosmos at his will. For his amusement, he has created a plethora of universes, each more intricate than its predecessor, starting with a world-system based on musical pitch (similar to the notions of Pythagoras). After the Star Maker grows tired of one universe, he destroys it and creates another. Our own expanding cosmos lies in the middle of a long series of creations. In future cycles, intelligent creatures will have powers far superior to our own, including the ability to explore all possible time paths. Ultimately, the universe will comprise a multidimensional labyrinth, embodying all conceivable worlds.

Stapledon's fiction contains the seeds of a modern-day mythos based on conscious life transcending an unstable universe. It is hard to find a better parable for the concept of human intelligence dealing with a cosmos that undergoes perpetual and total renewal.

Everlasting Life

What if it were indeed the case that the Big Crunch would mean the end of everything, including intelligent life? Even if it were absolutely impossible for the human race to survive the Big Crunch, there is still some chance that the subjective lifetime of our species would be infinite. We may not exist in bodily form forever, but perhaps we would exist as thinking entities for an infinite sequence of thoughts.

In an open universe, Freeman Dyson, of the Institute for Advanced Study at Princeton, has speculated that intelligent life might exist eternally, as heat death were approached, by continuing to lower its operating temperature. Humans would simply learn to think slower and use less energy. Currently, we operate at 300 degrees Kelvin, but if we took twice as long to think each thought, half as much energy per thought would be used, and we could run at 150 degrees Kelvin. As space continued to cool, we would just need to slow down our thoughts more and more. Theoretically, by using Dyson's method, there would be no limit to how long our race might last.

This method of indefinite survival might work fine in an open cosmos, but in a closed universe, cosmic cataclysm would come way before heat death, so this option wouldn't exist. Other methods of life-extension would need to be utilized as the Big Crunch was approached.

Barrow and Tipler have devised a method for determining the odds that intelligent life would survive forever, at least according to its own internal clock. In their work, they start off with the controversial assumption that the human mind is akin to a computer. (Many scientists, such as Penrose in *The Emperor's New Mind*, would beg to differ here.) Typically, a computer follows a program to gather data together in coded form. Each step of this computer program constitutes a "state" of the system. Now, the question of survivability can be expressed in the query: Does there exist, in our universe, a computer program that passes through an infinite number of distinct states?

Suppose such an eternal coding of data were possible. Then, Barrow and Tipler argue, by converting from one form of "hardware" (corporeal structure) to another, and by utilizing different energy sources, the human race could presumably survive forever, albeit no longer in the bodies we know. Thus, upon world's end, in a manner strangely reminiscent of Stapledon's speculated transformations, we would abandon our fleshy shells in favor of surroundings more suitable for Big Crunch conditions. Ultimately, we would become one with the bath of electrons, positrons and radiation that constitutes the twilight universe, and shape it to suit our needs. Under these rather peculiar circumstances, we would continue to process information for an infinite number of steps; hence, by our own reckoning, we'd live forever (some might ask, though, if that really would be living).

For this perpetual existence to be meaningful, Barrow and Tipler assert, there mustn't be a repetition of processing states. Following the same line of reasoning that leads to the idea of Poincaré recurrence, if the number of *distinct* computational states were finite, then there necessarily would be such reiteration.

Barrow and Tipler believe that such a situation would be bad for human morale, and hope that it would be avoided at all costs. Eternal return, they maintain, would remove the possibility of infinite progress, and hence would render life meaningless. Therefore, the conditions for infinite life must include not only an unlimited number of processing steps, but also the lack of duplication of states.

Finally, these physicists ponder the ultimate source of energy, as the universe is reduced to a point; for without energy, eternal storage of data would be impossible. To generate this power, future beings would learn how to tap into the gravitational forces of the universe itself, particularly the pushes and pulls caused by the chaotic fluctuations of the late-stage cosmos. Many theorists believe that conditions close to the omega point would be highly turbulent. Near the end of the universe, enormous gravitational tidal forces (stretching and squeezing motions) would likely be present. Therefore, Barrow and Tipler speculate that life could maintain itself indefinitely, by exploiting—in a manner similar to present-day experimental use of tidal power from the oceans—the gravitational energy of such turbulence. Of course, such a scheme is pure conjecture.

Another model for "eternal life" was recently proposed by Russian physicist, Andrei Linde. In contrast to Barrow and Tipler, Linde imagines that the Big Crunch conditions would be inordinately hostile for life. Once the universe has collapsed past a certain point, he asserts, random quantum fluctuations would damp out all meaningful collection of information. Therefore, another means of survival would need to be found.

The method developed by Linde involves sending information through a wormhole into another universe spawned from ours. Before describing Linde's notion in more detail, let's consider the speculative theories of Linde, Hoyle, Gott, and others, that there may be much more to our universe than what the eye (or dark matter detectors) can possibly see. Perhaps, as these innovative models assert, all of observable space represents merely a minute part of a greater cosmic entity.

OTHER COSMOS

In infinite space many civilizations are bound to exist, among them societies that may be wiser and more "successful" than ours. I support the cosmological hypothesis which states that the development of the universe is repeated in its basic characteristics an infinite number of times. Yet we should not minimize our sacred endeavors in this world, where, like faint glimmers in the dark, we have emerged for a moment from the nothingness of dark unconsciousness into material existence.
-Andrei Sakharov, 1975 Nobel Peace Prize Lecture

Bubble Universes

The ancients thought of Earth as the center of the cosmos; they couldn't imagine mundane worlds beside our own. The invention of the telescope changed all that, revealing that the planets and so-called fixed stars are just as central and tangible as Earth. In our own century, using increasingly powerful devices, we have extended the postulated limits of space even farther. We now believe that the universe consists of an enormous quantity of galaxies, arranged into clusters and superclusters. With each innovation our universe has seemingly gotten bigger and bigger.

In the past few years, many scientists have begun to talk about another tremendous extension of scope: the idea that our own universe is but one of many. In these recent theories, expanding universes lie beyond observation, created in a similar manner to our own. Thus even if our own universe died, it would not necessarily constitute the end of all creation; life might survive elsewhere.

Surprisingly, the granddaddy of all of these multiuniverse models is Fred Hoyle's steady-state cosmology. Hoyle (along with Jayant Narlikar) has been working on his theory of the universe for years, revising it, as necessary, when new evidence has disproved particular elements. Even the discovery of the cosmic microwave background and the COBE findings, thought by the overwhelming majority of researchers to be proof for the Big Bang model, hasn't discouraged him. It certainly isn't easy to be spokesman for a minority view, and Hoyle has held up his banner without flinching.

Hoyle's main philosophical difficulty with the standard model is the idea of the universe coming into being all at once. In his original steady-state theory, continuous creation of matter, in various centers around the universe, would replace a solitary Big Bang as the explanation for the Hubble expansion. This early approach was shown to be wrong by the findings of Penzias and Wilson, which demonstrated that the universe is bathed in the radiation produced by a universal explosion.

The new steady-state approach grants that the microwave background was generated during an era in which the observable universe was highly concentrated. However, to keep the idea of a perfect cosmological principle (stating that the cosmos must have the same general appearance for all times), it is assumed that the Big Bang was but one of many explosions, spread out about evenly across a larger space that is permanent. So, if one looks at the "big picture," the *entire universe* (I'll call this the "mega-universe"), in contrast to our own observable subsection, would appear to remain roughly the same for all times, with occasional explosions here and there (of which ours would be one).

An interesting version of this approach, called the *bubble universe model*, was developed in the early 1980s by Richard Gott. Gott postulates, in his theory, that the mega-universe consists of a continuously expanding de Sitter space, with no initial singularity. Recall that de Sitter space, an isotropic solution of Einstein's equation with either a cosmological constant or a nonzero vacuum energy, has the property of exponential expansion of its characteristic scale factor, allowing it to appear always the same. From what would this de Sitter space arise? Gott imagines that the mega-universe would continually generate itself, with its geometry creating its radiation, and its radiation molding its geometry.

Because in quantum theory there is no such thing as a perfect vacuum, small fluctuations would appear in this space. These would arise like bubbles in a vast expanding pool. In theory, because the mega-universe would be ever-growing, an infinite number of these bubbles could be accommodated. In a manner similar to the steady-state hypothesis, the enlargement of space would pull old bubbles apart, perpetually making way for new arrivals.

Each of these bubbles, once formed, would begin to expand outward at approximately the speed of light, and would constitute an independent Friedmann cosmology. Our own universe, rather than originating as a singularity, stemmed, according to Gott, from such a bubble. The fiery conditions of bubble production, rather than the force of motion from a single point, would have produced all observable Big Bang effects, including the cosmic microwave background and the current density ratio of hydrogen to helium in the universe.

What would happen if two bubble universes were to crash? The results, Gott asserts, would probably be disastrous. A "burst of very hot radiation coming from a small angular diameter disk in the sky" would be enough to obliterate the Earth instantly. Fortunately, Gott concludes, the chances of such a collision occurring within the next millennium are less than 1 in 15 million. So millennialists, don't worry.

Eternal Inflation

Linde's eternal inflation model, also proposed in the 1980s (and recently deemed, by Linde and Arthur Mezhlumian, the "stationary universe"), provides an alternative "steady-state-type" paradigm, differing from Gott's theory mainly in its incorporation of the inflationary universe approach. It is an extension of another of Linde's ideas, known as chaotic inflation, in which an inflationary era is precipitated by a random set of primordial field fluctuations in the very early universe.

According to the eternal inflation scenario, the universe would start off in a vacuum state of some sort. Because no vacuum is perfectly empty, it would be sprinkled here and there with small fluctuations. Some of these fluctuations would be of the right sort to form the seeds for regions of extremely rapid expansion.

Linde argues that there would be an infinite number of these seeded domains, each experiencing the sudden flash of an inflationary era, followed by a more humdrum period of standard Friedmann-type growth. Therefore, as in the case of Gott's model, there would be an infinite number of island universes, forever out of communication with each other, each one evolving in its own manner as part of a extraordinarily vast mega-universe.

The "eternal" part comes in when one considers what would happen to our own universe in its later stages of development. Quite likely, large numbers of ultradense black holes would be formed from the myriads of massive dying stars. If the universe is flat or open, there is a good chance that our region of space would end up as part of a huge collapsed region—a sort of giant black hole. If the universe instead is closed, then all spatial points will eventually recongregate in a Big Crunch; therefore, everything will some day lie within a mammoth black hole. Either way, Linde mournfully asserts, the cosmos and all life within would be doomed to die out. Without a doubt, the human race, along with all other intelligent creatures, would be ultimately fated for extinction.

There would be, however, one consolation in all this tragedy, according to Linde. The center of this contracted region of space, being highly dense, would be an ideal spawning ground for small quantum field fluctuations. This dynamic froth would yield, in turn, a large number of self-reproducing inflationary domains, each appearing like an exponentially growing universe, attached to our own by a wormhole. This process would happen eternally, with one generation of universes generating the next, hence the name "eternal inflation."

So, if we managed somehow to travel through one of these wormholes (almost certainly impossible, because of the sinister black hole conditions), or at least to send information about our race (somewhat more likely), to one of the

Figure 25. Andrei Linde. (Courtesy of Andrei Linde.)

newly formed baby universes, there would be the distinct possibility of a new breed of intelligent life, cognizant of our own, forming in the time beyond the Big Crunch. We might even produce informational portraits of individual human beings and transmit this data to a new cosmos. Thus, along with Barrow and Tipler's proposed scheme, this would be another way to ensure that our species—or at least intelligent life in general—doesn't completely die out (this is all quite theoretical; there may not be wormholes at all).

Tipler, however, points out one perceived major flaw with Linde's mechanism for "eternal life," namely, that repetition of states would be inevitable. In a 1992 article, entitled, appropriately enough, "The Ultimate Fate of Life in Universes which Undergo Inflation," Tipler examines the long-term consequences of trying to perpetuate our kind by sending data to spawned universes, and finds that this process would necessarily lead to eternal return in the form of Poincaré cycles.

Whether the return of all human beings to their original forms would be a blessing or a curse is, as we've seen, a matter of opinion. For Tipler, eternal return would be a curse. He states:

> In a progressive universe, it is at least possible that our efforts make a permanent, and ultimately important, contribution on a cosmic scale. In (an) Eternal Return cosmology, our civilization is doomed to sink back into 'the nothingness of dark unconsciousness.'

To drill in his point, Tipler humorously concludes his paper with an echo of Thomas Hobbes: "In summary, life in an eternal chaotic inflationary universe would ultimately be poor, solitary, nasty, brutish, and—by comparison with the infinite age of the universe—short."[1]

The Best of All Possible Worlds

The question of whether or not our own universe represents the best of all possible worlds for the existence and sustenance of intelligent life has been well debated in recent years, especially since the 1986 publication of Barrow and Tipler's comprehensive (and controversial) text, *The Anthropic Cosmological Principle*.

The anthropic principle is a sort of litmus test that can be applied to models of the universe. If a theorist proposed that a particular solution to Einstein's equation represents our own universe, and if it were subsequently found that the cosmology produced conditions unacceptable to the formation of conscious beings, then one would have sufficient grounds with which to reject the proposal. In other words, because our own universe contains living thinking creatures, it must have originated from the subset of all possible worlds that was able to produce living, thinking creatures.

This may seem rather obvious; a credible universe model should, one might think, predict the existence of life. For some cosmologists, therefore, the anthropic principle is trivial. For example, Roger Penrose feels that very little would be gained by applying this criterion; a wide range of initial conditions could still lead to a cosmos with life. Yet many other researchers, including, most notably, Hawking, Barrow, and Tipler, feel that the restrictions imposed would be considerable, enough to narrow down considerably the range of feasible universe models. Ultimately, these theorists would like to use the anthropic principle to prove why the constants of physics have the values they have.

How would the anthropic principle work to perform this selection? Let's consider an example of a cosmological theory that would *not* produce life. Suppose that in a particular model, the binding energy of hydrogen to oxygen were such that stable water molecules would be impossible to produce. Any water formed in this universe, let's say, would be completely unstable and would disintegrate in a few milliseconds. Then, certainly, this universe model—with its altered set of physical constants—would be inappropriate, because life (we assume) needs water to exist. Following the anthropic principle, this model would be removed from consideration.

Other Cosmos

The most important potential application of the anthropic principle lies in the consideration of bubble universes, eternal inflation models, and other schemes involving parallel cosmological developments. In each of these cases, anthropic considerations might be used to assess the likelihood that life might develop in other parts of the multiuniverse.

We assume, based on our own limited experiences, that our own universe is the one that is best suited for intelligent beings, but perhaps we are wrong. Maybe somewhere in the vast array of bubble cosmos, there are universes that are overflowing with life of all forms, much more so than ours? Perhaps creatures in these other worlds would be surprised to hear that our own universe is even inhabited? Thus, even though the anthropic principle is a fascinating and powerful conjecture, one clearly cannot make any conclusive statements about life in the cosmos until *all* worlds are explored.

Natural Selection

A rather bizarre variation of the anthropic principle notion, utilizing a cosmological theory of evolution, was proposed in 1992 by Lee Smolin of Syracuse University. Smolin uses a unique measure of universal "survival of the fittest," along with theories (such as Linde's) that black holes would spawn baby universes, to answer the question of why the known physical constants possess the particular values they have (rather than different values).

Basically, Smolin's model determines the probability that a given universe contains particular physical characteristics. He uses, as a measure of this, the number of collapsed stars within the universe. In his Darwinian approach, he shows that the propensity of a universe to produce black holes in cosmology might be compared to the procreation rate of a species in biology.

According to some rather speculative theories, the interiors of black holes harbor the seeds of "baby universes." These baby universes, once born, would develop and grow in a higher dimensional space. Therefore, "mama universes" that contain more collapsed stars would be more successful reproductively, and the characteristics of these black-hole-filled universes would be passed on with more frequency. In this manner, the features of universes with many black holes would be more likely to spread from generation to generation, and the features of those with few would be more likely to die out.

When Darwin proposed his theory of biological natural selection, he noted a wide variation of characteristics among species. Darwin couldn't fully explain how these discrepancies come into being, but today we know that they are due to genetic mutations. Radiation and other effects cause DNA patterns to become altered, and, through "survival of the fittest," successful variations become passed down more often. Eventually, environmentally favored features win out.

In Smolin's analogous scheme, small random variations of physical constants would occur whenever a baby universe was generated by a black hole. Successful "mutations" of this sort—leading to larger numbers of black holes in successors—would be passed down with greater frequency. Unsuccessful mutations would lead to evolutionary dead ends. Hence, following this logic, a universe with characteristics such that no black holes were formed during its lifetime would represent the cosmological equivalent of a dodo bird or moa; it would rapidly become extinct.

How would a universe's physical constants affect how many black holes it harbors? Essentially, the answer lies with the process of nucleosynthesis. If, on the one hand, these constants were such that a large number of massive stars would be produced from hydrogen, then naturally many black holes would accumulate—produced from the remnants of these stars after they die.

On the other hand, suppose that conditions in a cosmos were such that the mass difference between neutrons and protons were much greater. One can readily calculate that this large mass discrepancy would make it virtually impossible for helium nuclei to form, and hence would significantly hamper the production of stars. Because fewer stars would form, fewer would die, and fewer black holes would be created. Finally, due to natural selection, such a universe, having generated almost no offspring, would die with few heirs with its features.

In short, Smolin's answer to the question of why the cosmos possesses particular physical constants is the fact that if these constants were much changed, the universe would have far fewer black holes. And if the universe contained fewer black holes, it would be a member of a dying evolutionary line. Because it is much more probable that our cosmos is from a thriving evolutionary line (with more family members), we must assume that the current physical parameters of the universe have approximately the same values that led to the reproductive success of the universe's ancestors. Thus, ultimately, according to Smolin, the existence of these physical constants is due to their associated potency.

The Meaning of It All

In reviewing Smolin's idea that the universe exists in order to generate lots and lots of black holes, I am reminded of a more fanciful idea by Kurt Vonnegut, who, in the *Sirens of Titan*, concludes that the entire history of our world is designed to enable an alien civilization to send a brief message across space. Humankind, in its ceaseless agitation for knowledge, is not content to accept that things are because they *are*; it must find an unmistakable purpose for everything under the sun.

Increasingly, physicists are stepping foot into the quagmire of universal explanation and are trying to determine, using experimentation and logic, the purpose and destiny of all creation. And like the antics of Charlie Chaplin's "Little Tramp," in his magnificent balletlike stumbling, these efforts are both awe inspiring and pathetic. Awe inspiring because we have gone so far in so little time, with so few failures of which to speak. We've managed to walk the high-wire of astrophysical measurement without falling on our face; we've gracefully executed a search that has produced remarkably consistent values for many universal parameters. And there is strong reason to believe that our poised performance will be rewarded with many tangible results; for example, we'll probably soon know the value of omega and whether or not the cosmos will eventually collapse.

The pathos arises because there are all too clear limits to even our most powerful efforts. We cannot travel back to the time of the Big Bang, let alone journey into the netherworld (or other cycles) that may have preceded the universal explosion. We cannot step beyond our own universe to see if there are others. And it is physically impossible to peer through the singularity of a black hole, to determine what lies beyond the edges of time and space. Thus, unfortunately, a great deal of what is currently debated in the halls of academia will remain mere speculation for a long time, but we hope not forever.

EPILOGUE

THE PASSION FOR DISCOVERY

Full circle we go, round and round, debating the universal questions over and over again. In our spinning carousel of inquiry, we reach out for the brass ring of our fate, hoping someday to grasp the ultimate prize: knowledge of the fate of the cosmos. Why do we try so hard to obtain this elusive treasure?

Alone among the creatures of the planet, the human race is blessed with a full sense of time and history, as well as the longing to understand its own past and future. Though this interpretation comes in different forms—some cultures view time as a cycle, while others perceive it as a straight line—what is shared is a desire to map out and comprehend human destiny. Therefore, we quest for our fate because it is in our very nature. We are self-aware beings that possess the supreme joy of discovery. Even if we sometimes travel in circles, we enjoy going along for the ride.

From Galileo's early earthbound device, to the modern digital camera and other optical instruments on board the orbiting Hubble, for the past few centuries the window to our fate has been in the form of a telescope. Even as you read this page, somewhere in the world light is being gathered from the sky, recorded, and then carefully analyzed in order to provide astronomers with a more accurate picture of our universal destiny. Is the universe open, closed, or flat? Will it someday collapse to a point, and, if so, how quickly? The data being collected by the latest generation of optical instruments will help us to resolve these monumental questions.

Within the next few decades, assuming that all goes well, scientists will present us with a detailed model of the future evolution of the universe. We will finally know whether or not the cosmos is closed in on itself like a great cosmic serpent. An ancient riddle will at last be resolved, and, having accomplished this task, the human race will likely move on to the next cosmic mystery. Such is the passion for discovery, the lot of our inquisitive species.

REFERENCES

Chapter One:

1. Fritjof Capra, *The Tao of Physics* (New York: Bantam Books, 1980), p. 232.
2. M. Oldfield Howey, *The Encircled Serpent: A Study of Serpent Symbolism in All Countries and Ages* (Philadelphia: David McKay, 1926), p. 17.
3. Seneca, *Physical Science in the Times of Nero: Being a Translation of the Quaestiones Naturales of Seneca*, translated by John Clarke (London: MacMillan, 1910), p. 153.
4. Jeremiah Ostriker, interviewed by Florence Helitzer for the *Intellectual Digest*, June 1973. Cited in *Reincarnation: The Phoenix Fire Mystery*, compiled and edited by Joseph Head and S. L. Cranston, (New York: Julian Press/Crown Publishers, 1977), p. 432.

Chapter Two:

1. Friedrich Nietszche, *The Will to Power*, translated by A. Ludovici (London: T. N. Foulis, 1910), p. 1066.

Chapter Four:

1. Marcia Bartusiak, *Thursday's Universe* (New York: Times Books, 1986), p. 211.
2. Andrei Linde, Private communication.
3. Andrei Linde, *Ibid*.

Chapter Seven:

1. David Leckrone and Ed Weiler, quoted in Vincent Kiernan "Hubble's Eye Can See Clearly Now," *New Scientist* (January 22, 1994): 4.

Chapter Eight:

1. Richard F. Mushotzky and John Mulchaey, quoted in Kathy Sawyer, "Astronomers Discover Mysterious Dark Matter," *Philadelphia Inquirer*, January 5, 1993, p. 1.
2. Phone conversation with John Mulchaey of the University of Maryland, April 28, 1994.

Chapter Nine:

1. Fredric Brown, "Experiment," in *The Best of Fredric Brown*, edited by Robert Bloch (New York: Ballantine Books, 1977), p. 277.
2. P. C. W. Davies, "The Physics of Time Asymmetry," in *The Study of Time III: Proceedings of the Third Conference of the International Society for the Study of Time*, edited by J. T. Fraser, N. Lawrence, and D. Park, (New York: Springer-Verlag, 1978), p. 99.
3. James Hartle, "Excess Baggage," In *Elementary Particles and the Universe*, edited by John H. Schwarz, (New York: Cambridge University Press, 1991), p. 7.
4. Stephen Hawking, *A Brief History of Time: From the Big Bang to Black Holes* (New York: Bantam Books, 1988), p. 150.

Chapter Ten:

1. John Gribbin, *The Omega Point: The Search for the Missing Mass and the Ultimate Fate of the Universe* (New York: Bantam Books, 1988), p. 2.
2. Edward Harrison, *Cosmology* (London: Cambridge University Press, 1961), p. 300.

Chapter Eleven:

1. Frank Tipler, "The Ultimate Fate of Life in Universes which Undergo Inflation," *Physics Letters B*, Vol. 286 (July 1992): 36.

RELATED READING

The following is a list of general and technical references for the reader who wishes to know more about the subject. References marked with an asterisk are of a more technical nature.

Introduction: The Farthest Supernova

Binkley, Sue, *The Clockwork Sparrow: Time, Clocks and Calendars in Biological Organisms* (Englewood Cliffs, New Jersey: Prentice Hall, 1990).

Goldsmith, Donald, *Supernova: The Violent Death of a Star* (New York: Oxford University Press, 1990).

Halpern, Paul, *Time Journeys: A Search for Cosmic Destiny and Meaning* (New York: McGraw-Hill, 1990).

Priestly, J. B., *Man and Time* (London: Bloomsbury Books, 1989).

Szamosi, Géza, *The Twin Dimensions: Inventing Time and Space* (New York: McGraw-Hill, 1986).

Chapter One: The Endless Dance of Shiva

Bhattacharjee, Siva Sadhan, *The Hindu Theory of Cosmology* (Calcutta: Bani Prakashani, 1978).

Cranston, Sylvia, and Williams, Carey, *Reincarnation: A New Horizon in Science, Religion and Society* (New York: Julian Press, 1984).

Eliade, Mircea, *Cosmos and History: The Myth of the Eternal Return* (New York: Harper and Row, 1959).

Halpern, Paul, *Time Journeys: A Search for Cosmic Destiny and Meaning* (New York: McGraw-Hill, 1990).

Henderson, Joseph, and Oakes, Maud, *The Wisdom of the Serpent: The Myths of Death, Rebirth and Resurrection* (New York: George Braziller, 1963).

Jaki, Stanley, *Science and Creation: From Eternal Cycles to an Oscillating Universe* (Edinburgh: Scottish Academic Press, 1986).

Martin, Eva, *Reincarnation: The Ring of Return* (New Hyde Park, New York: University Books, 1964).

Mundkur, Balaji, *The Cult of the Serpent: An Interdisciplinary Survey of Its Manifestations and Origins* (Albany, New York: State University of New York Press, 1983).

Reyna, Ruth, "Metaphysics of Time in Indian Philosophy," In *Time in Science and Philosophy*, edited by Jiri Zeman (New York: Elsevier, 1971), pp. 227–239.

Warner, Rex, *The Greek Philosophers* (New York: New American Library, 1986).

Waterfield, Robin, *Before Eureka: The Presocratics and their Science* (New York: St. Martin's Press, 1989).

Chapter Two: Eternal Return: The Mirror of Chance

Boltzmann, Ludwig, *Theoretical Physics and Philosophical Problems* (Boston: D. Reidel Publishing, 1974).

Broda, Engelbert, *Ludwig Boltzmann: Man, Physicist, Philosopher* (Woodbridge, Connecticut: Ox Bow Press, 1983).

Brush, Stephen, *Kinetic Theory* (New York: Pergamon Press, 1966).

Brush, Stephen, *The Temperature of History: Phases of Science and Culture in the Nineteenth Century* (New York: Burt Franklin, 1978).

Halpern, Paul, *Time Journeys: A Search for Cosmic Destiny and Meaning* (New York: McGraw-Hill, 1990).

Lea, F. A., *The Tragic Philosopher: A Study of Friedrich Nietzsche* (New York: Philosophical Library, 1957).

Nietzsche, Friedrich, *The Will to Power*, translated by A. Ludovici (London, T. N. Foulis, 1910).

Schacht, Richard, *Nietzsche* (London: Routledge and Kegan, 1983).

Schlegel, Richard, "Time and Entropy," in *Time in Science and Philosophy*, edited by Jiri Zeman (New York: Elsevier, 1971), pp. 27–35.

Stambaugh, Joan, *The Problem of Time in Nietzsche* (Lewisburg, Pennsylvania: Bucknell University Press, 1987).

Whitrow, G. J., *The Natural Philosophy of Time* (Oxford: Clarendon Press, 1980).

Chapter Three: The Expanding Heavens

Barrow, John, and Silk, Joseph, *The Left Hand of Creation: The Origin and Evolution of the Expanding Universe* (New York: Basic Books, 1983).

Bartusiak, Marcia, *Thursday's Universe* (New York: Times Books, 1986).

Friedman, Herbert, *The Astronomer's Universe: Stars, Galaxies and Cosmos* (New York: Ballantine Books, 1990).

Harrison, Edward, *Cosmology* (London: Cambridge University Press, 1981).

Lightman, Alan, *Ancient Light: Our Changing View of the Universe* (Cambridge, Massachusetts: Harvard University Press, 1991).

Parker, Barry, *The Vindication of the Big Bang: Breakthroughs and Barriers* (New York: Plenum, 1993).

Silk, Joseph, *The Big Bang: The Creation and Evolution of the Universe* (San Francisco: W. H. Freeman, 1980).

Chapter Four: Full of Sound and Fury

*Albrecht, Andreas, and Steinhardt, Paul, "Cosmology for Grand Unified Theories with Radiatively Induced Symmetry Breaking," *Physical Review Letters* Vol 48 (April 1982): 1220.

Barrow, John, and Silk, Joseph, *The Left Hand of Creation: The Origin and Evolution of the Expanding Universe* (New York: Basic Books, 1983).

*Guth, Alan, "Inflationary Universe: A Possible Solution to the Horizon and Flatness Problems" *Physical Review, D15* Vol. 23 (January 1981): 347.

*Linde, Andrei, "A New Inflationary Universe Scenario: A Possible Solution of the Horizon, Flatness, Homogeneity, Isotropy and Primordial Monopole Problems" *Physics Letters, B* Vol. 108 (February 1982): 389.

*Linde, Andrei, "Chaotic Inflation" *Physics Letters, B* Vol. 177 (September 1983): 129.

Parker, Barry, *Einstein's dream: The Search for a Unified Theory of the Universe* (New York: Plenum, 1986).

Parker, Barry, *The Vindication of the Big Bang: Breakthroughs and Barriers* (New York: Plenum, 1993).

Weinberg, Steven, *The First Three Minutes: A Modern View of the Origin of the Universe* (New York: Basic Books, 1977).

Chapter Five: Mapping Our Fate

Einstein, Albert, *Relativity: The Special and General Theory* (New York: Crown Publishers, 1961).

Related Reading

Gribbin, John, *The Omega Point: The Search for the Missing Mass and the Ultimate Fate of the Universe* (New York: Bantam Books, 1988).

Isham, Jamal, *The Fate of the Universe* (New York: Cambridge University Press, 1983).

*Misner, Charles, Thorne, Kip, and Wheeler, John, *Gravitation* (San Francisco: W. H. Freeman, 1973).

Pais, Abraham, *Subtle Is the Lord: The Science and Life of Albert Einstein* (New York: Oxford University Press, 1982).

*Peebles, P. J. E., *Principles of Physical Cosmology* (Princeton: Princeton University Press, 1993).

*Peebles, P. J. E., "Tests of Cosmological Models Constrained by Inflation." *The Astrophysical Journal*, Vol. 284 (May 1984): 439.

*Weinberg, Steven, *Gravitation and Cosmology: Principles and Applications of the General Theory of Relativity* (New York: John Wiley and Sons, 1972).

Chapter Six: The Shape of Creation

Abbott, Edwin, *Flatland* (New York: Dover, 1953).

Burger, Dionys, *Sphereland* (New York: Thomas Y. Crowell, 1965).

Field, George, and Goldsmith, Donald, *The Space Telescope: Eyes Above the Atmosphere* (New York: Contemporary Books, 1989).

Isham, Jamal, *The Fate of the Universe* (New York: Cambridge University Press, 1983).

Kline, Morris, *Mathematics in Western Culture* (New York: Oxford University Press, 1953).

Matz, Lloyd, and Weaver, Jefferson Hane, *The Unfolding Universe: A Stellar Journey* (New York: Plenum, 1989).

Parker, Barry, *Invisible Matter and the Fate of the Universe* (New York: Plenum, 1989).

Rucker, Rudy, *The Fourth Dimension* (Boston: Houghton Mifflin, 1984).

Chapter Seven: Galactic Speeding Tickets

Bartusiak, Marcia, *Thursday's Universe* (New York: Times Books, 1986).

Cohen, Nathan, and Goldsmith, Donald, *Mysteries of the Milky Way* (New York: Contemporary Books, 1991).

Cook, William, and Schrof, Joannie, "Journey to the Beginning of Time," *U. S. News and World Report* (March 26, 1990): p. 52.

*Ellis, Richard, "Prospects for Measuring the Deceleration Parameter," in *Observational Tests of Cosmological Inflation*, edited by T. Shanks, A. J. Banday,

Related Reading

R. S. Ellis, C. S. Frenk, and A. W. Wolfendale (Boston: Kluwer Academic Publishers, 1991), pp. 243–249.

*Feast, M. W., "The Local Distance Scale: How Reliable Is It?" in *Observational Tests of Cosmological Inflation*, edited by T. Shanks et al. (Boston: Kluwer Academic Publishers, 1991), pp. 147–160.

Field, George, and Goldsmith, Donald, *The Space Telescope: Eyes Above the Atmosphere* (New York: Contemporary Books, 1989).

Flamsteed, Sam, "How Big is the Universe?" *Discover* (November 1992): 49.

Gott, J. Richard III, Gunn, James E., Schramm, David N., and Tinsley, Beatrice, "Will the Universe Expand Forever?" *Scientific American* (March 1976): 62.

*Guiderdoni, B., "High-Redshift Tests of Omega," in *Observational Tests of Cosmological Inflation*, edited by T. Shanks et al. (Boston: Kluwer Academic Publishers, 1991), pp. 217–231.

*Hodge, Paul W., "The Extragalactic Distance Scale," in *The Early Universe: Reprints*, edited by Edward Kolb and Michael Turner (New York: Addison-Wesley, 1988), pp. 41–53.

Hodge, Paul W., "The Extragalactic Distance Scale: Agreement at Last?" *Sky and Telescope* (October 1993): 16.

*Huchra, John, "On Contemporary Observational Cosmology: When Did It All Begin?" in *Creation and the End of Days: Judaism and Scientific Cosmology*, edited by David Novak and Norbert Samuelson (New York: University Press of America, 1985), pp. 27–39.

*Huchra, John, "The Cosmological Distance Scale," in *The Early Universe: Reprints*, edited by Edward Kolb and Michael Turner (New York: Addison-Wesley, 1988), pp. 55–61.

Kiernan, Vincent, "Hubble Repair Boosts Space Station's Prospects," *New Scientist* (December 1993): 6.

*Pritchet, C. J., "Novae and the Distance Scale," in *Observational Tests of Cosmological Inflation*, edited by T. Shanks et al. (Boston: Kluwer Academic Publishers, 1991), pp. 199–203

*Rowan-Robinson, Michael, "Distances to Virgo and Beyond," in *Observational Tests of Cosmological Inflation*, edited by T. Shanks et al. (Boston: Kluwer Academic Publishers, 1991), pp. 161–171.

*Sandage, Allan, and Tammann, Gustav, "The Dynamical Parameters of the Expanding Universe as They Constrain Homogeneous World Models with and without a Cosmological Constant," in *The Early Universe: Reprints*, edited by Edward Kolb and Michael Turner (New York: Addison-Wesley, 1988), pp. 15–35.

Schwarzschild, Bertram, "Supernova Distance Measurements Suggest an Older, Larger Universe," *Physics Today* (November 1992): 17.

Related Reading

*Shanks, T., Tanvir, N., Doel, P., Dunlop, C., Myers, R., Major, J., Redfern, M., Devaney, N., and O'Kane, P., "A High Resolution, Ground Based Observation of a Virgo Galaxy," in *Observational Tests of Cosmological Inflation*, edited by T. Shanks et al. (Boston: Kluwer Academic Publishers, 1991), pp. 205–210.

*Tammann, Gustav, "Observation Status of H_0." In *Observational Tests of Cosmological Inflation*, edited by T. Shanks et al. (Boston: Kluwer Academic Publishers, 1991), pp. 179–185.

*Weinberg, Steven, *Gravitation and Cosmology: Principles and Applications of the General Theory of Relativity* (New York: John Wiley and Sons, 1972).

Chapter Eight: The Search for Missing Matter

Bartusiak, Marcia, *Through A Universe Darkly: A Cosmic Tale of Ancient Ethers, Dark Matter and the Fate of the Universe* (New York: HarperCollins, 1993).

Cowen, R., "Dark Matter: MACHOS in Milky Way's Halo?" *Science News* (September 1993): 199.

Croswell, Ken, "Universe Will Expand Forever," *New Scientist* (December 18, 1993): 14.

Davis, M., Efstathiou, G., Frenk, C. S., and White, S. D. M., "The End of Cold Dark Matter?" *Nature* Vol. 356 (April 1992): 489.

Ferris, Timothy, *Coming of Age in the Milky Way* (New York: William Morrow, 1988).

*Geller, Margaret, "The Universe: Always Room for More?" In *Creation and the End of Days: Judaism and Scientific Cosmology*, edited by David Novak and Norbert Samuelson (New York: University Press of America, 1985), pp. 79–95.

*Hodges, Hardy M., "Mirror Baryons as the Dark Matter," *Physical Review, D15* Vol. 47 (January 1993): 456.

Krauss, Lawrence, *The Search for the Fifth Essence: Dark Matter in the Universe* (New York: Basic Books, 1989).

Parker, Barry, *Invisible Matter and the Fate of the Universe* (New York: Plenum, 1989).

Parker, Barry, "The Missing Mass Mystery," *Astronomy* (November 1984): 6.

Redfern, Martin, "Galactic Cannibalism Has Made Milky Way Big and Fat." *New Scientist* (June 5, 1993): 14.

Riordan, Michael, and Schramm, David, *The Shadows of Creation: Dark Matter and the Structure of the Universe* (New York: W. H. Freeman, 1991).

Tayler, Roger, *The Hidden Universe* (New York: Ellis Horwood, 1991).

Trefil, James, *The Dark Side of the Universe* (New York: Scribners, 1988).

Related Reading

*Trimble, Virginia, "Existence and Nature of Dark Matter in the Universe." In *The Early Universe: Reprints*, edited by Edward Kolb and Michael Turner (New York: Addison-Wesley, 1988), pp. 67–105.

Tucker, Wallace, and Tucker, Karen, *The Dark Matter* (New York: Morrow, 1988).

Tyson, Anthony, "Mapping Dark Matter with Gravitational Lenses," *Physics Today* (June 1992): 24.

*White, Simon D. M., "Dynamical Estimates of Omega from Galaxy Clustering," in *Observational Tests of Cosmological Inflation*, edited by T. Shanks et al. (Boston: Kluwer Academic Publishers, 1991), pp. 279–291.

*White, Simon D. M., Navarro, Julio F., Evrard, August E., and Frenk, Carlos S., "The Baryon Content of Galaxy Clusters: A Challenge to Cosmological Orthodoxy," *Nature*, Vol. 366 (December 1993): 429.

Chapter Nine: Reverse Performance

Asimov, Isaac, "Nightfall," In *The Best of Isaac Asimov* (Garden City, New York: Doubleday, 1974), pp. 9–18.

Asimov, Isaac, *The Collapsing Universe* (New York: Walker, 1977).

Borges, Jorge Luis, *The Aleph and Other Stories*, edited and translated by Norman Thomas di Giovanni (New York, E. P. Dutton, 1970).

Brown, Fredric, "Experiment," in *The Best of Fredric Brown*, edited by Robert Bloch (New York: Ballantine Books, 1977), pp. 277–278.

Costa de Beauregard, O. "Time in Relativity Theory: Arguments for a Philosophy of Being," in *The Voices of Time*, edited by J. T. Fraser (New York: George Braziller, 1966), pp. 256–272.

Davies, Paul, *God and the New Physics* (New York: Simon and Schuster, 1983).

Dick, Philip K., *Counter-Clock World* (New York: Berkley Medallion Books, 1967).

Dick, Philip K., *Ubik* (New York: Vintage Books, 1969).

Ellis, George, *Before the Beginning: Cosmology Explained* (New York: Boyers/Bowerdean, 1993).

Gribbin, John, "Could Time Run Backwards?" *New Scientist* (August 1993): 14.

Hawking, Stephen, *A Brief History of Time: From the Big Bang to Black Holes* (New York: Bantam Books, 1988).

*Hawking, Stephen, Laflamme, R., and Lyons, G. W., "Origin of Time Asymmetry," *Physical Review*, D15 Vol. 47 (June, 1993): 5342.

Morris, Richard, *Time's Arrows* (New York: Simon and Schuster, 1984).

*Penrose, Roger, "Time-Asymmetry and Quantum Gravity," in *Quantum Gravity 2*, edited by Jamal Isham, Roger Penrose, and Dennis Sciama (New York: Oxford University Press, 1981), pp. 105–138.

Schlegel, Richard, "Time and Entropy," In *Time in Science and Philosophy*, edited by Jiri Zeman (New York: Elsevier, 1971), pp. 27–35.

Watson, Ian, *The Very Slow Time Machine* (New York: Ace Books, 1979).

Zee, Anthony, "Time Reversal," *Discover* (October 1992): 96.

Zelazny, Roger, "Divine Madness," in *Trips in Time*, edited by Robert Silverberg (New York: Thomas Nelson, 1977), pp. 17–28.

Chapter Ten: After the Crunch

Asimov, Isaac, "The Last Question," in *The Best of Isaac Asimov* (Garden City, New York: Doubleday, 1974), pp. 157–169.

Bailey, James, *Pilgrims Through Space and Time: Trends and Patterns in Utopian and Science Fiction* (Westport, Connecticut: Greenwood Press, 1972).

Barrow, John, and Tipler, Frank, *The Anthropic Cosmological Principle* (New York: Oxford University Press, 1986).

Carr, Donald E., *The Eternal Return* (Garden City, New York: Doubleday, 1968).

*Dyson, Freeman, "Time without End: Physics and Biology in an Open Universe," *Reviews of Modern Physics*, Vol. 51 (July 1979): 447.

Gribbin, John, *The Omega Point: The Search for the Missing Mass and the Ultimate Fate of the Universe* (New York: Bantam Books, 1988).

Gribbin, John, *Unveiling the Edge of Time: Black Holes, White Holes, Wormholes* (New York: Harmony, 1992).

*Guth, Alan and Sher, Marc, "The Impossibility of a Bouncing Universe," *Nature*, Vol. 302 (April 1983): 505.

Halpern, Paul, *Cosmic Wormholes: The Search for Interstellar Shortcuts* (New York: Dutton, 1992).

Harrison, Edward, *Cosmology* (London: Cambridge University Press, 1981).

Hirsch, Eli, *The Concept of Identity* (New York: Oxford University Press, 1982).

*Linde, Andrei, "Life after Inflation." *Physics Letters, B* Vol. 211 (August 1988): 29.

*Morris, Michael, and Thorne, Kip, "Wormholes in Spacetime and Their Use for Interstellar Travel: A Tool for Teaching General Relativity," *American Journal of Physics* Vol. 56 (August 1988): 395.

Moskowitz, Sam, *Explorers of the Infinite* (New York: World, 1963).

Penrose, Roger, *The Emperor's New Mind: Concerning Computers, Minds, and the Laws of Physics* (New York: Oxford University Press, 1989).

*Petrosian, Vahe, "Phase Transitions and Dynamics of the Universe." *Nature*, Vol. 298 (August 1982): 805.

*Rozental, I. L., *Big Bang, Big Bounce: How Particles and Fields Drive Cosmic Evolution* (New York: Springer-Verlag, 1987).

*Saslaw, William C., "Black Holes and Structure in an Oscillating Universe." *Nature*, Vol. 350 (March 1991): 43.

Sorabji, Richard, *Time, Creation and the Continuum: Theories in Antiquity and the Early Middle Ages* (Ithaca, New York: Cornell University Press, 1983).

Stapledon, Olaf, *Last and First Men and Star Maker: two science-fiction novels* (New York: Dover, 1968).

Teilhard de Chardin, Pierre, *The Future of Man* (New York: Harper and Row, 1959).

*Tipler, Frank, "General Relativity, Thermodynamics and the Poincaré Cycle," *Nature*, Vol. 280 (July 1979): 203.

*Tolman, R. C., *Relativity, Thermodynamics and Cosmology* (Oxford: Oxford University Press, 1934).

Chapter Eleven: Other Cosmos

Breuer, Reinhard, *The Anthropic Principle: Man as the Focal Point of Nature* (Boston: Birkhauser, 1991).

Close, Frank, *Apocalypse When?* (New York: William Morrow, 1988).

*Gott, J. Richard III, "The Very Early Universe," in *Creation and the End of Days: Judaism and Scientific Cosmology*, edited by David Novak and Norbert Samuelson (New York: University Press of America, 1985), pp. 41–56.

Gribbin, John, "Evolution of the Universe by Natural Selection?" *New Scientist* (February 1, 1992): 14.

*Linde, Andrei, "Chaotic Inflation," *Physics Letters, B* Vol. 129 (September 1983): 177.

*Linde, Andrei, "Life after Inflation," *Physics Letters, B* Vol. 211 (August 1988): 29.

*Linde, Andrei, "Life after Inflation and the Cosmological Constant Problem," *Physics Letter,s B* Vol. 227 (August 1989): 352.

*Linde, Andrei, and Mezhlumian, Arthur, "Stationary Universe," *Physics Letters, B* Vol. 307 (August 1993): 25.

Sakharov, Andrei, *Alarm and Hope* (New York: Knopf, 1978).

*Smolin, Lee, "Did the Universe Evolve?" *Classical and Quantum Gravity*, Vol. 9 (January 1992): 173.

*Tipler, Frank, "The Ultimate Fate of Life in Universes which Undergo Inflation," *Physics Letters, B* Vol. 286 (July 1992): 36.

Velan, A. Karel, *The MultiUniverse Cosmos: The First Complete Story of the Origin of the Universe* (New York: Plenum, 1992).

*Vilenkin, Alexander, "Did the Universe Have a Beginning?" *Physical Review, D15* Vol. 46 (September 1992): 2355.

INDEX

Abbott, Edwin, 165–166
Albrecht, Andreas, 126
Aleph, The (Borges), 228
Alpher, Ralph, 98
Andromeda galaxy, 71, 92, 94–95, 183
Anthropic Cosmological Principle, The (Barrow and Tipler), 273
Anthropic principle, 273–275
Aristotle, 78–79
Arrow of time, 227–228, 231
Asimov, Isaac, 221, 223, 225, 258
Astrology, 23, 26
Astroparticle physics, 106–108
Atomism, 50, 55–56, 64–67
Avasarpini, 17
Aztecs, 15, 21–22, 37, 223

Babylonians, 15, 22–26, 28, 31, 37, 71, 78–79, 149
Background radiation, 102–103, 114–117, 128–130, 154, 197, 268, 270

Bacon, Francis, 41
Bailes, Matthew, 202
Barrow, John, 257, 264–266, 272–274
Bartusiak, Marcia, 106
Beauregard, O. Costa de, 228
Berossus, 24–26
Bible, 15, 40, 81, 88
Big Bang, 4–6, 43, 96–103, 129–130, 200, 237–254, 270
Big Bounce, 5, 261
Big Crunch, 5, 154, 237, 242–245, 248–251, 255, 262–266
Black dwarfs, 60–62
Black holes, 60–63, 144, 154, 204, 210, 239, 254–256, 271, 275–278
Boltzmann, Ludwig, 55–56, 62–68, 70
Bolyai, John, 161–163
Bondi, Hermann, 97
Borges, Jorge Luis, 228
Bosons, 106–108, 122, 216–217, 230

Index

Brahma, 12, 37–39, 69, 75
Brahman, 36, 39
Brief History of Time, A (Hawking), 42, 237
Brown dwarfs, 201–202
Brown, Frederic, 234
Browne, Ian, 192
Bubble universe, 42, 269
Buddhism, 34
Burger, Dionys, 166
Burstein, David, 213–214

Calendar Round, 20
Canary Islands, 6, 188
Capra, Fritjof, 13, 81
Carnot, Sadi, 57–58
Catalogue of Nebulae, 87
Cepheid variable stars, 70, 89–94, 182–187, 195
Chaldeans, 23
Chaotic inflation, 126–127, 270
Chaplin, Charlie, 278
Christianity, 33, 40–41, 79
Chromatic aberration, 82
Ciardullo, Robin, 183
Cicero, 26
Clausius, Rudolf, 57–58, 62
Closed universe model, 149, 155, 168, 186, 249, 259–264
Cold dark matter, 215–216
Collapse of universe: *see* Big Crunch
Conservation of energy, 48–49, 53
Copernicus, Nicholas, 79
Cosmic background: *see* Background radiation
Cosmic Background Explorer (COBE), 116, 197, 128–130, 154, 239, 268

Cosmological constant, 146
Cosmological distance ladder, 89, 181–183
Cosmological Principle, 97, 114, 145, 147, 151, 169, 268, 273
Cosmology
  Babylonian, 28
  Greek, 28–31
  Hindu, 13, 36–40, 69
  Newtonian, 83
  physical, 75–76
  scientific, 88
  steady state, 97–100
Counter-Clock World (Dick), 233
Crab Nebula, 2, 60
Cronin, J.W., 230
Cycles
  Babylonian, 23–26, 28, 79
  Central American, 20, 22, 37
  Chinese, 16, 22, 37
  Egyptian, 18–19
  Greek, 26–33, 37, 79
  Indian, 11–13, 17, 34–40, 79
Cyclical serpent: *see* Ourobouros
Cyclical time, 16

Dahomeans, 15
Dalton, John, 56
Dance of Shiva, 11–14, 33, 40
Dark matter, 110–111, 197, 199–200, 205–218, 266
Darwin, Charles, 41, 276
Davies, Paul, 81, 235
De Sitter, Willem, 146–147
De Vaucouleurs, Gérard, 183, 186
Deceleration parameter, 179–183, 187–189, 193, 197
Democritus, 54
Dick, Philip K., 233

Index

Dicke, Robert, 101–102, 123, 152
Distance ladder, 89, 181–183
Divine Madness (Zelazny), 234
Doppler effect 71, 94–96, 115, 181, 191
Dunne, John William, 228
Dyson, Freeman, 264

Eddington, Arthur, 91, 96, 247
Egyptians, 15, 18, 19, 78, 218
Einstein, Albert, 42–43, 83, 96, 131, 135–139, 142–147, 160, 239
Einstein's equation, 142–147, 151, 154, 217, 262, 269, 274
Einstein Ring, 192
Elamites, 16, 244
Electrons, 28, 107, 110–111, 117, 119, 123, 230, 265
Elementary particles, 28, 54, 60, 69, 105–110, 121–122, 126
Empedocles, 29–30, 47
Enlightenment, 41
Eternal inflation, 270–273
Eternal life, 266, 273
Eternal return, 25, 42–43, 45–50, 67, 69, 71, 259–261, 265, 273
Euclid, 88, 160, 162
Expansion of universe: *see* Big Bang; Hubble expansion
Experiment (Brown), 234–235
Experiment with Time, An (Dunne), 228
Extrasolar planets, 202–203

Fermions, 106–108, 216–217
Fisher, Richard, 185
Fitch, Val 230

Flat universe model 148–150, 168, 218
Flatland (Abbott), 165–167, 171
Frail, Dale, 203
Friedman, Wendy, 182
Friedmann, Alexander, 146–147

Galileo, 76–83, 279
Gamow, George, 98, 102
Gauss, Karl Friedrich, 88, 160–162
Gell-Mann, Murray, 235–236
General relativity, 42, 135, 139–144, 160, 163, 170, 217, 238, 251
Glashow, Sheldon, 106
God and the New Physics (Davies), 81
Gold, Thomas, 97, 226, 252
Goldhaber, Gerson, 6, 188
Gott, Richard, 269, 270
Grand unification theory (GUT), 121–124
Gravitational lenses, 189, 191–192, 211–212
Great Attractor, 206–207
Great Year, 19, 22, 25–26, 31, 149
Gribbin, John, 249–250
Gulliver's Travels (Swift), 27
Guth, Alan, 121–126

H-theorem, 65–68
Hale, George Ellery, 92
Harmony of the spheres, 28–29
Hartle, James, 235–236
Hawking, Stephen, 42, 229, 236–245, 249, 251, 274
Hawkins, Isabel, 116–117
Helmholtz, Hermann, 62–63
Henriksen, Mark, 213–214

Heraclitus, 29–30, 47
Herschel, William, 84, 86–89, 183
Hewitt, Jacqueline, 191–192
Hinduism, 11–14, 17, 33–40, 43, 67, 69, 75, 79, 88, 149, 225, 245
Horus Apollo, 18
Hot dark matter, 215–216
Hoyle, Fred, 4, 97, 100, 266, 268
Hubble, Edwin, 70–71, 92–96, 129, 154, 177
Hubble expansion, 109, 119, 128, 146, 171, 226, 231, 249, 268
Hubble parameter, 150, 180, 185–187, 191–200, 213
Hubble Space Telescope, 192–196, 202
Hutton, James, 41

Inflation
  chaotic, 126–127, 270
  eternal, 270–273
  new, 126–127
  old, 119–125
Inflationary universe, 119–128, 130, 151, 153, 238–239, 270–273
Infrared Astronomical Satellite (IRAS), 202
Intellectual Digest, 43
Invisible matter: see Dark matter
Isaac Newton telescope, 6, 188
Ishtar, 15
Israel, Werner, 255

Jacoby, George, 183
Jainas, 17, 34, 244
Jesus, 24, 34
Judaism, 40

Kalpa, 37–40, 67, 149
Kaluza, Theodor, 169–171
Kaluza–Klein model, 171
Kaon decay, 230
Kirshner, Robert, 188
Kirzhnits, David, 124
Klein, Oskar, 170–171
Kuhn, Thomas, 88

Laflamme, Raymond, 237–238, 242–243
Last Question, The (Asimov), 258
Law of entropy, 58–59, 61–63, 65, 67, 226
Law of inertia, 51
Leavitt, Henrietta, 89–91, 182
Leckrone, David, 197
Lehár, Joseph, 191
Lemaître, Abbe Georges, 95–96
Lin, Douglas, 209
Linde, Andrei, 124, 126–127, 266, 270–273, 275
Linear time, 40
Lippershey, Hans, 76
Lobatchevsky, Nikolai, 161–163
Local Group, 71, 95, 181, 183
Lyell, Charles, 41
Lyne, Andrew, 202
Lyubimov, V. A., 215

Mach, Ernst, 55–56
Machetto, Duccio, 195
Madore, Barry, 182
Magellanic Clouds, 90–91, 209–210
Mahayuga, 37–38
Mamon, Gary, 213–214
Mass-to-luminosity ratio, 200–201

Index

Massive compact halo objects (MACHOs), 210–211
Maxwell, James Clerk, 56, 122
Mayans, 20–21, 37
Mechanical determinism, 52
Mendeleev, Dmitry, 56
Mesopotamia, 22–23, 25
Messier catalogue, 92
Meyer, David, 116
Mezhlumian, Arthur, 270
Microwave background: see Background radiation
Milky Way, 70–71, 86–89, 91–94, 113, 116, 172, 181–183, 208–210
Milne, E. A., 63
Missing mass: see Dark matter
Morris, Michael, 262
Mount Wilson Observatory, 92
Mulchaey, John, 212–214
Multi-Element Radio-Linked Interferometer Network (MERLIN), 192
Mushotzky, Richard, 199, 212–214

Naimittik Pralaya, 39
Narlikar, Jayant, 268
Neutron star, 60, 202
New General Catalogue, 87
New Testament, 41
Newton, Isaac, 50–54, 57, 76, 82–83, 139, 146
Newtonian mechanics, 50, 65
Nietzsche, Friedrich, 42–50, 55, 67, 69, 71, 259
Nightfall (Asimov), 223–225
No boundary condition, 238, 240–245, 249, 251
No-return theorem, 261

Non-Euclidean geometry, 160–162
Numerology, 27–28

Old Testament, 40
Oliveira-Neto, G., 241
Omega point, 248–251, 255–257, 261, 266
Omega Point, The (Gribbin), 250
On the Construction of the Heavens (Herschel), 86
On the Revolutions of Celestial Bodies (Copernicus), 79
Oort, Jan, 205, 207
Open universe model, 147–150, 154, 168, 174, 258, 264
Oscillating universe model, 42, 101, 245–246, 250–255, 258–260
Osiris, 19
Ostriker, Jeremiah, 43
Ostwald, Wilhelm, 55–56
Ourobouros, 16–17, 19, 35, 42, 106, 235, 244–246

Page, Don, 237–238, 242
Panagia, Nino, 195
Parallax, 181–182
Peebles, Phillip James, 152
Pennypacker, Carl, 6, 188
Penrose, Roger, 237, 239, 251, 254, 264, 274
Penzias, Arno, 101–103, 114, 129, 154, 268
Perfect cosmological principle, 97, 268
Perlmutter, Saul, 6, 188
Petrosian, Vahe, 256
Pierce, Michael, 187

Pius, Emperor Antonius, 32
Planetary nebulae, 87, 182–183
Planets, 23–26, 29, 31, 77–80, 83–86, 200–203, 208, 222, 258, 267
Plato, 29–32, 226
Pliny the Elder, 26
Poincaré, Henri, 42, 67–69, 71, 259–261, 265, 273
Poincaré recurrence, 67–69, 259–261, 265, 273
Poisson, Eric, 255
Prakrta Pralaya, 39
Principle of Equivalence, 140–142
Pritchet, Chris, 182
Pulsar, 202–203
Puranas, 11, 37, 39, 43
Pythagoras, 26–31, 263

Quantum field theory, 121
Quantum gravity, 261
Quantum mechanics, 28, 56, 107, 170, 227, 229, 256
Quantum theory, 52, 56, 106–107, 229–230, 236, 254, 261, 269
Quasars, 188, 190–192, 196
Quetzalcoatl, 15

Readhead, Tony, 192
Reading the Mind of God (Trefil), 81
Recombination era, 117–118
Recurrence, 7, 19, 25, 30, 46–50, 64, 66–69, 71, 259, 261, 265
Recurrence paradox, 66–67
Reincarnation, 15–16, 18, 31, 36, 260

Relativity: *see* General Relativity; Special Relativity
Renaissance, 41, 48
Riemann, Bernhard, 162
Robertson, H. P., 151
Roentgen Satellite (ROSAT), 212–214
Romans, 24–26, 32–33
Roth, Katherine, 116–117
Rowan-Robinson, Michael, 207
RR Lyrae, 182–183
Rubin, Vera, 206–207

Saha, Abhijit, 195
Sandage, Allan, 183–187, 194–196, 199
Sanskrit, 14, 35–36
Second law of thermodynamics, 58–59, 61–63, 65, 67, 226
Seneca, 24–26
Serpents, 11–12, 14–19, 22, 34, 45, 47, 96, 217, 219, 244, 280
Serpent symbolism, 15–16
Shanks, Tom, 187
Shapley, Harlow, 182
Shemar, Setnam, 202
Shiva, 11–14, 33, 35, 39–40, 47
Sikkema, A. E., 255
Sirens of Titan (Vonnegut), 278
Slipher, Vesto, 95
Smolin, Lee, 275–277
Smoot, George, 115–117, 129–130, 154
Snakes: *see* Serpents
Special relativity, 137–139
Spencer, Herbert, 41
Sphereland (Burger), 166–167, 169, 172
Spin, 106–107, 159, 185, 216–217

Index

Spiral nebulae 87, 91–92, 94
St. Augustine, 41
Standard candles, 6, 173–175, 182–183, 188–189, 194
Standard model, 103, 105, 113, 154, 236, 268
Stapledon, Olaf, 263
Star Maker, The (Stapledon), 263
Starobinsky, Alexei, 121
Steady state universe model, 97–98, 226, 268–269
Steinhardt, Paul, 126
Stellar Movements and the Structure of the Universe (Eddington), 91
Stoics, 31–33, 35
Structure of Scientific Revolutions, The (Kuhn), 88
Superconducting Supercollider (SSC), 108
Supercooling, 119, 124, 126–127
Supernovas, 1–3, 6–7, 60, 98–99, 187–189, 194–196, 223
Supersymmetry, 216–217
Susa, 17, 244
Swift, Jonathan, 27

Tammann, Gustav, 183–187, 195, 199
Tao of Physics, The (Capra), 13, 81
Taoism, 16
Teilhard de Chardin, Pierre, 249
Telescopes: *see* Hubble Space Telescope; Isaac Newton Telescope
Thomson, William, 56
Thorne, Kip, 262
Thus Spoke Zarathustra (Nietzsche), 45–47

Time irreversibility, 54, 57
Time reversibility, 52–54, 57
Time's arrow, 57, 61, 65–66, 231
Tipler, Frank, 257, 260–261, 264–266, 272–274
Tolman, Richard, 252–254
Trefil, James, 81
Tryon, Ed, 254
Tully, Brent, 185
Tully–Fisher relation, 185
Turner, Ed, 191
Tyson, Anthony, 211
Tzolkin, 20

Ubik (Dick), 233
Universe
  baby, 276
  bubble, 268–270
  Friedmann–Robertson–Walker, 151
  inflationary, 119–128, 130, 151, 153, 238–239, 270–273
  oscillating, 245–246
  steady state, 97–98, 100
Upanishads, 36, 225
Utsarpini, 17

Van den Bergh, Sidney, 183
Vedas, 12, 35–37, 88, 218, 225
Very Large Array (VLA), 191
Very Slow Time Machine, The (Watson), 233–234
Virgo cluster, 94, 185, 187
Vishnu, 12, 37, 40
Vonnegut, Kurt, 277

Walker, A. G., 151
Watson, Ian, 233
Wave function, 229–230, 236

Weiler, Ed, 197
Wheel of Samsara, 36, 40
Wheeler, John, 257
White dwarf, 60
Wilkinson, Peter, 192
Wilson, Robert, 101–103, 114, 129, 154, 268
Wolszczan, Alexander, 202–203
Wormholes, 262, 266, 271

Yin-yang, 16, 22
Yuga, 37–38

Zarathustra, 45–48, 50
Zelazny, Roger, 234
Zeno, 31
Zermelo, Ernst, 67, 69
Zoroastrianism, 47
Zwicky, Fritz, 96–97, 195, 205–207